THE WATER WIZARD

The Extraordinary Power of Natural Water

Viktor Schauberger

translated & edited by
Callum Coats

GATEWAY BOOKS
BATH, U.K.

First published in 1998
by GATEWAY BOOKS
The Hollies, Wellow,
Bath, BA2 8QJ, UK

(c) The Heirs of Viktor and Walter Schauberger
(c) Introduction by Callum Coats

Distributed in the USA by
ACCESS PUBLISHERS NETWORK
6893 Sullivan Road,
Grawn, MI 49637

All rights reserved. No part of this book may be reproduced or transmitted in any form or by any means, electronic or mechanical, including photocopying or any information retrieval system, without permission in writing from the Publishers.

Set in 10.5pt on 12.5pt Palatino, by
Character Graphics (Taunton) Ltd
Printed and bound in Great Britain by
Redwood Books of Trowbridge.

Cover design by Synergie, Bristol

British Library Cataloguing-in Publication Data:
A catalogue record for this book is
available from the British Library

ISBN: 1 85860 049 9

Contents

Foreword	v
A Brief Introduction to the Natural Eco-Technological Theories of Viktor Schauberger	1
The Nature Of Water	**15**
The Cancerous Decay of Organisms	19
The Substance – Water	19
Concerning Processes of *Ur*-Creation, Evolution and Metabolism	22
High-Frequency Water	26
The Natural Reconversion of Seawater into Fresh Water	33
Fire under Water	38
Notes on the Secret of Water	41
The Production of Fuels	42
The Difference between Energising Substances and Fuels	44
The Quantitative and Qualitative Deterioration of Water	**45**
The Deterioration of Water	45
The Sterilisation of Water	46
Consequences of Chlorination of Water	47
The Consequences of Contemporary Water-Purification Processes	48
An Experiment	50
Water Supply and the Mechanical Production of Drinking Water	**53**
Water Supply	53
The Consequences of Producing Drinking Water by purely Mechanical Means	57
Deep-Sea Water	58
The Conduction of the Earth's Blood	**61**
The Double-spiral Flow Pipe	62
The Pulsation of Water	66
Healing Water for Human, Beast and Soil	69
Temperature and the Movement of Water & other unpublished texts on River Engineering	**81**
River Regulation - My Visit to the Technical University for Agricultural Science	82

Turbulence - Concerning the Movement of Water and its Conformity with Natural Law	89
"Temperature and the Movement of Water"	94
Temperature Gradients - Full & Half Hydrological Cycles	94
The Groundwater Table	96
The Drainage of Water	99
Basic Principles of River Regulation	102
The Interrelationship between Groundwater & Agriculture	109

Fundamental Principles of River Regulation & Status of Temperature in Flowing Water — 106

Turbulent Phenomena in Flowing Water	106
Temperature Gradient, Riverbed-slope and River Bend Formation	107
The Influence of the Geographical Situation and the Rotation of the Earth	112
The General Tasks of River Regulation	115
The Regulation of Temperature Gradient	119
The Movement of Temperature in Mass-Concrete Dam Walls	122
Expert Opinion of Professor Philipp Forchheimer	131

The Natural Movement of Water over the Earth's Surface — 135

Tractive Force Considered	159
Concerning Rivers and Water	165
The Transport of Sediment: Timber, Ore and Other Materials Heavier than Water	167

The Rhine and the Danube — 170

The Problem of the Danube Regulation	170
The Rhine Battle	176
Energy-Bodies	179

The Dr. Ehrenberger Affair — 183

The Learned Scientist and the Star in the Hailstone — 193

Appendix: Patent Applications — 201

Glossary — 212

Index — 216

Foreword

It was a Swedish engineer and anthroposophist, Olof Alexandersson, who wrote the first popular introduction to the radical ideas of Viktor Schauberger. I came across this attractive little book in 1979 and had it translated into English. *Living Water* is now in its eighth printing and has inspired many to go on to Callum Coats' in-depth study of Schauberger's ideas, *Living Energies,* which was published in 1996. My friendship with Callum goes back to 1981 when he confided in me his wish to write a definitive work on Viktor Schauberger. Callum had met Viktor's son, Walter Schauberger, in 1977 and was to spend three years studying with Walter at his Pythogoras-Keppler System Institute in Lauffen, in the Saltzkammergut near Salzburg. During that time, Callum was given access to all Viktor's writings.

Viktor Schauberger did not start seriously to write about his ideas and his discoveries until the age of 44, when he acquired a distinguished sponsor in Professor Philipp Forchheimer. As Callum describes later in this volume, Forchheimer, a world famous hydrologist, had been asked by the Austrian Government to report on Schauberger's controversial log flumes, which transported large amounts of timber from inaccessible locations without damage. He was so impressed with Schauberger's discoveries that he asked him to write a paper which was published in 1930 in *Die Wasserwirtschaft,* the Austrian Journal of Hydrology. This paper attracted the attention of the President of the Austrian Academy of Science, Professor Wilhelm Exner, and resulted in a commission to write a more detailed study of his theories for that same magazine under the title *Temperature and the Movement of Water.*

Schauberger's ideas flew completely in the face of conventional ideas of hydrology and water management and, as a result, gained him many enemies in scientific circles. The reason Viktor developed the strong feelings about orthodox scientific research that you will read in this and subsequent volumes was partly to defend himself from their attacks, and partly out of his despair at witnessing the ongoing destruction of the natural environment by their blind and uncaring technologies. It was this despair that motivated him to write his only book, *Our Senseless Toil - the Cause of the*

World Crisis. It was published at a time of severe depression, when many were worried about the future.

After Forchheimer died, Schauberger found another ally in Professor Werner Zimmermann who encouraged Viktor in 1935-1936 to write about the damage being wrought to the great rivers, the Rhine and the Danube, in a small 'new thought' magazine *Tau*. After Schauberger's death, two magazines published further collections of Schauberger's writings: *Implosion* was started by a student and collaborator of Viktor's, and published a number of his articles in the 1960s. *Mensch und Technik* in the 1970s published articles by and about Viktor Schauberger for the more free-thinking scientific community.

Callum Coats has skilfully woven together these articles, together with correspondence with other scientists, friends and officials of one kind or another, into a fascinating tapestry which gives a true and very readable account of Schauberger's impassiond campaign to alert the world to the dangers of the prevailing scientific dogma. Unfortunately, not much has changed, and Schauberger's vision of how humanity must work cooperatively with Nature if we are to have a future, is perhaps more relevant than ever.

Callum arranged this massive amount of material into a large volume, *Eco-Technology*. In considering this for publication, we realised that it would be much more accessible in several volumes, arranged by theme. This first one, *The Water Wizard,* is devoted to Schauberger's ideas about water and rivers. The second, *Nature as Teacher,* concerns the wider implications of his ideas on water and energy. The third, *The Fertile Earth,* describes the way trees transform energy, and the processes of fertilisation of the soil. The final volume, *The Energy Revolution,* gathers together the discussion and description of Schauberger's appliances for purifying and energising water and for producing vast amounts of virtually free energy. Together with *Living Energies,* the *Eco-technology* series give a complete account of the vision and genius of one of the founders of the present ecological movement, and are an inspiration for all those who wish to see our precious Earth saved from extinction by short-sightedness and greed, and the emergence of a new partnership with bountiful Nature.

Alick Bartholomew, Wellow, December 1997

A Brief Introduction to the Natural Eco-Technological Theories of Viktor Schauberger

Viktor Schauberger (30 June 1885 – 25 September 1958) was born in Austria of a long line of foresters stretching back some four hundred years. He developed a gift for accurate and intuitive observation so great that he was able to perceive the natural energies and other phenomena occurring in Nature, which are still unrecognised by orthodox science. Refusing to attend University at the age of 18, to the fury of his father, Viktor Schauberger left home and spent a long period alone in the high, remote forest, contemplating, pondering and observing any subtle energetic processes taking place in Nature's laboratory, where they were still undisturbed by human hand. During this period he developed very profound and radical theories, later to be confirmed practically, concerning water, the energies inherent in it and its desired natural form of motion. These eventually earned him the name of 'The Water Wizard'.

For the whole of his life he fought a running and often acrimonious battle with academia and its institutions, since his theories in the main were diametrically opposed to the so-called established facts of science. His practical demonstration of them always functioned as he had theorised, however, for he had come to understand the true inner workings of Nature and was able to emulate them.

Viktor Schauberger's theories afford new insights into the naturally correct or 'naturalesque' management of water. This encompasses its proper handling, storage, and conduction by means that promote its self-purification, the retention and enhancement of its natural energies and health. In this book, the close interrelationship between water and the forest (as a water-producer - not a water consumer) is examined. The problem of soil salinity and how this comes about through over-exposure of the soil to the radiance of the Sun through deforestation and faulty agricultural practices, are also addressed. Indications are given as to how these may be avoided and overcome, due to Viktor Schauberger's radical and fundamentally new understanding of the coming into being and functioning of the groundwater table in relation to soil temperature.

As a natural organism, water is formed and functions according to

Nature's laws and geometry, the latter exhibiting none of the elements of the straight line, circle and point, the basis of modern mechanical and technological artefacts. Reflecting Nature's principal constant, namely that of continuous change and transformation, the vortex epitomises this form of open, fluid and flexible motion. Through his study of the vortices occurring naturally in flowing water and in the air in the form of cyclones and tornadoes, Viktor Schauberger developed his theories of implosion. It was through the research and development of these theories that he was able to produce spring-quality water and generate considerable energies in and with water and air.

In listing some of his accomplishments one could not do better than to quote from his book, *Our Senseless Toil*, written in 1933:

"It is possible to regulate watercourses over any given distance without embankment works; to transport timber and other materials, even when heavier than water, for example ore, stones, etc., down the centre of such water-courses; to raise the height of the water table in the surrounding countryside and to endow the water with all those elements necessary for the prevailing vegetation."

"Furthermore it is possible in this way to render timber and other such materials non-inflammable and rot resistant; to produce drinking and spa-water for man, beast and soil of any desired composition and performance artificially, but in the way that it occurs in Nature; to raise water in a vertical pipe without pumping devices; to produce any amount of electricity and radiant energy almost without cost; to raise soil quality and to heal cancer, tuberculosis and a variety of nervous disorders."

"... the practical implementation of this ... would without doubt signify a complete reorientation in all areas of science and technology. By application of these new found laws, I have already constructed fairly large installations in the spheres of log-rafting and river regulation, which as is known, have functioned faultlessly for a decade, and which today still present insoluble enigmas to the various scientific disciplines concerned."

Water and its vital interaction with the forest was Viktor's principal preoccupation. He viewed water as a living entity, the 'Blood of Mother-Earth', which is born in the womb of the forest. Our mechanistic, materialistic and extremely superficial way of looking at things, however, prevents us from considering water to be anything other than inorganic, i.e. supposedly without life but, while apparently having no life itself, can nevertheless miraculously create life in all its forms. Life is movement and is epitomised by water, which is in a constant state of motion and transformation, both externally and internally. In confirmation of this fact, water is able to combine with more substances than any other molecule and, flowing as water, sap and blood, is the creator of the myriad life-forms on this planet. How then could it ever be construed as life-less in accordance with the chemist's clinical view of water, defined as the inorganic substance H_2O? This short description is a gross misrepresentation. As the fundamental basis of all life, water is itself a living entity and should be treated as such. Failure to do so quickly transforms it into an enemy, rather than the nurturer and furtherer of all life that it should be.

> *"This civilisation is the work of man, who high-handedly and ignorant of the true workings of Nature, has created a world without meaning or foundation, which now threatens to destroy him, for through his behaviour and his activities, he, who should be her master, has disturbed Nature's inherent unity."*

Apart from the more familiar categories of water, there are, according to Viktor Schauberger, as many varieties of water as there are animals and plants. Were water merely the sterile, distilled H_2O as claimed by science, it would be poisonous to all living things. H_2O or 'juvenile water' is sterile, distilled water and devoid of any so-called 'impurities'. It has no developed character and qualities. As a young, immature, growing entity, it grasps like a baby at everything within reach. It absorbs the characteristics and properties of whatever it comes into contact with or has attracted to itself in order to grow to maturity. This 'everything' - the so-called 'impurities'- takes the form of trace elements, minerals, salts and even smells! Were we to drink pure H_2O constantly, it would quickly leach out all our store of minerals and trace elements, debilitating and ultimately killing us. Like a growing child, juvenile water takes and does not give. Only when mature, i.e. when suitably enriched with raw materials, is it in a position to give, to dispense itself freely and willingly, thus enabling the rest of life to develop. Before the birth of water, there was no life.

But what is this marvellous, colourless, tasteless and odourless substance, which quenches our thirst like no other liquid? Did we but truly understand the essential nature of water - a living substance - we would not treat it so churlishly, but would care for it as if our lives depended on it, which undoubtedly they do.

"The Upholder of the Cycles which supports the whole of Life, is water. In every drop of water dwells the Godhead, whom we all serve; there also dwells Life, the Soul of the 'First' substance - Water - whose boundaries and banks are the capillaries that guide it and in which it circulates."

"More energy is encapsulated in every drop of good spring water than an average-sized powerstation is presently able to produce.

Indeed in accordance with the famous Hasenöhrl-Einstein equation $E = mc^2$, in 1 gram of substance, or 1 cubic centimetre of water, 25 million kilowatt hours of energy are stored!

Water is a being that has life and death. With incorrect, ignorant handling, however, it becomes diseased, imparting this condition to all other organisms, vegetable, animal and human alike, causing their eventual physical decay and death, and in the case of human beings, their moral, mental and spiritual deterioration as well. From this it can be seen just how vital it is, that water should be handled and stored in such a way as to avert such pernicious repercussions.

"Science views the blood-building and character-influencing ur-organism[1] - 'water' merely as a chemical compound and provides millions of people with a liquid prepared from this point of view, which is everything but healthy water."

But what does modern, de-naturised civilisation care, as long as it receives a suitably hygienised, clear liquid to shower, wash its dishes, clothes and cars. Once down the plug-hole in company with all manner of toxic chemicals and detergents, all is comfortingly out of sight and out of mind.

> "Our primeval Mother Earth is an organism that no science in the world can rationalise. Everything on her that crawls and flies is dependent upon Her and all must hopelessly perish if that Earth dies that feeds us."

Although the chlorination of drinking and household water-supplies ostensibly removes the threat of water-borne diseases, it does so, however, to the detriment of the consumer. In its function of water steriliser or disinfectant, chlorine eradicates all types of bacteria, beneficial and harmful alike. More importantly, however, it also disinfects the blood (about 80% water) or sap (ditto) and in doing so kills off or seriously weakens many of the immunity-enhancing micro-organisms resident in the body of those constantly forced to consume it. This eventually impairs their immune systems to such a degree that they are no longer able to eject viruses, germs and cancer cells, to which the respective host-bodies ultimately falls victim.

---

[1] In Viktor Schauberger's writings in German, the prefix 'Ur' is often separated from the rest of the word by a hyphen, e.g. 'Ur-sache' in lieu of 'Ursache', when normally it would be joined. By this he intends to place a particular emphasis on the prefix, thus endowing it with a more profound meaning than the merely superficial. This prefix belongs not only to the German language, but in former times also to the English, a usage which has now lapsed. According to the Oxford English Dictionary, 'ur' denotes 'primitive', 'original', 'earliest', giving such examples as 'ur-Shakespeare' or 'ur-origin'. This begins to get to the root of Viktor's use of it and the deeper significance he placed upon it. If one expands upon the interpretation given in the Oxford English Dictionary, then the concepts of 'primordial', 'primeval', 'primal', 'fundamental', 'elementary', 'of first principle', come to mind, which further encompass such meanings as: - pertaining to the first age of the world, or of anything ancient; - pertaining to or existing from the earliest beginnings;- constituting the earliest beginning or starting point;- from which something else is derived, developed or depends;- applying to parts or structures in their earliest or rudimentary stage; - the first or earliest formed in the course of growth. To this can be added the concept of an 'ur-condition' or 'ur-state' of extremely high potential or potency, a latent evolutionary ripeness, which given the correct impulse can unloose all of Nature's innate creative forces. In the English text, therefore, the prefix 'ur' will also be used wherever it occurs in the original German and the reader is asked to bear the above in mind when reading what follows. - Ed.

The appearance of AIDS, therefore, and the enormous increase in all forms of disease, cancer in particular, would have come as no surprise to Viktor Schauberger. Apart from the other inevitable disturbances to the ecology and the environment occasioned by humanity's unthinking activities, he foresaw it all as early as 1933.

"For a person who lives 100 years in the future, the present comes as no surprise."

Apart from other factors (some cannot be defined quantitatively), encompassing such aspects as turbidity (opaqueness), impurity, and quality, the most crucial factor affecting the health and energy of water is temperature.

As a liquid, the behaviour of water differs from all other fluids. The latter become consistently and steadily denser with cooling, water reaches its densest state at a temperature of +4°C (+39.2°F), below which it grows less dense. In contrast, water's behaviour is anomalous, because it reaches its greatest density at a temperature of +4°C (+39.2°F). This is the so-called 'anomaly point', or the point of water's anomalous expansion, which is decisive in this regard and has a major influence on its **quality**. Below this temperature it once more expands. This highest state of density is synonymous with its highest energy content, a factor to be taken carefully into account, since energy can also be equated with life or life-force. Therefore if water's health, energy and life-force are to be maintained at the highest possible level, then certain precautions must be taken, which will be addressed later.

Conceived in the cool, dark cradle of the virgin forest, water ripens and matures as it slowly mounts from the depths. On its upward way it gathers to itself trace elements and minerals. Only when it is ripe, and not before, will it emerge from the womb of the Earth as a spring. As a true spring, in contrast to a seepage spring, this has a water temperature of about +4°C (+39.2°F). Here in the cool, diffused light of the forest it begins its long, life-giving cycle as a sparkling, lively, translucent stream, bubbling, gurgling, whirling and gyrating as it wends its way valleywards. In its natural, self-cooling, spiralling, convoluting motion, water is able to maintain its vital inner energies, health and purity. In this way it acts as the conveyor of all the necessary minerals, trace elements and other subtle energies to the surrounding environment. Naturally flowing water seeks to flow in darkness or in the diffused light of the forest, thus avoiding the damaging direct light of the sun. Under these conditions, even when cascading down in torrents, a stream will only rarely overflow its banks. Due to its correct natural motion, the faster it flows, the greater its carrying capacity and scouring ability and the more it deepens its bed. This is due to the formation of in-winding, longitudinal, clockwise - anti-clockwise alternating spiral vortices down the central axis of the current, which constantly cool and re-cool the water, maintaining it at a healthy temperature and leading to a faster, more laminar, spiral flow.

To protect itself from harmful effects of excess heat, water shields itself from the Sun with over-hanging vegetation, for with increasing heat and light it begins to lose its vitality and health, its capacity to enliven and animate the environment through which it passes. Ultimately becoming a broad river, the water becomes more turbid, the content of small-grain sediment and silt increasing as it warms up, its flow becoming slower and more sluggish. However, even this turbidity plays an important role, because it protects the deeper water-strata from the heating effect of the sun. Being in a denser state, the colder bottom-strata retain the power to shift sediment of larger grain-size (pebbles, gravel, etc.) from the centre of the watercourse. In this way the danger of flooding is reduced to a minimum. The spiral, vortical motion mentioned earlier, which eventually led Viktor Schauberger to the formation of his theories concerning 'implosion', creates the conditions, where the germination of harmful bacteria is inhibited and the water remains disease-free.

Another of its life-giving properties is its high specific heat - lowest at +37.5°C (+99.5°F). The term "specific heat" refers to the capacity and rapidity of a body to absorb or release heat. With a relatively small input of heat fluids with a high specific heat warm up less rapidly than those with a lower specific heat. How strange then, and how remarkable, that the lowest specific heat of this "inorganic" substance - water - lies but 0.5°C (0.9°F) above the normal +37°C (+98.6°F) blood temperature of the most highly evolved of Nature's creatures - human beings. This property of water to resist rapid thermal change enables us, with blood composed of 80% water, to survive under large variations of temperature. Pure accident so we are told, or is it by clever, symbiotic design?! However, we are used to thinking about temperature in gross terms (car engines operate at temperatures of 1,000°C (1,832°F) or so and many industrial processes employ extremely high temperatures). Despite the fact that we begin to feel unwell if our temperature rises by as little as 0.5°C (0.9°F), we fail to see that non-mechanical, organic life and health are based on very subtle differences in temperature. When our body temperature is +37°C (98.6°F) we do not have a 'temperature' as such. We are healthy and in a state that Viktor Schauberger called 'indifferent' or 'temperature-less'. Just as good water is the preserver of our proper bodily temperature, our anomaly point of greatest health and energy, so too does it preserve this planet as a habitat for our continuing existence. Water has the capacity to retain large amounts of heat and were there no water vapour in the atmosphere, this world of ours would be an icy-cold, barren wasteland. Water in all its forms and qualities is thus the mediator of all life and deserving of the highest focus of our esteem.

"To Be or Not to Be: In Nature all life is a question of the minutest, but extremely precisely graduated differences in the particular thermal motion within every single body, which continually changes in rhythm with the processes of pulsation."

"This unique law, which manifests itself throughout Nature's vastness and unity and expresses itself in every creature and organism, is the 'law of ceaseless cycles' that in every organism is linked to a certain time span and a particular tempo."

"The slightest disturbance of this harmony can lead to the most disastrous consequences for the major life forms."

*"In order to preserve this state of equilibrium, it is vital that the characteristic **inner** temperature of each of the millions of micro-organisms contained in the macro-organisms be maintained."*

The No. 1 enemy of water is excess heat or over-exposure to the Sun's rays. It is a well-known fact that oxygen is present in all processes of organic growth and decay. Whether its energies are harnessed for either one or the other is to a very great extent, if not wholly, dependent on the temperature of the water as itself or in the form of blood or sap. As long as the water-temperature is below +9°C (+48.2°F), its oxygen content remains passive. Under such conditions the oxygen assists in the building up of beneficial, high-grade micro-organisms and other organic life. However, if the water temperature rises above this level, then the oxygen becomes increasingly active and aggressive. This aggressiveness increases as the temperature rises, promoting the propagation of pathogenic bacteria, which, when drunk with the water, infest the organism of the drinker.

"Thus the development of micro-organisms and the opportunities for their propagation are simply a result of the condition in which the respective sickening macro-organism finds itself, which will be attacked by these parasites and which eventually must fall victim to them if its inner climatic conditions are no longer strictly regulated."

But this aggressiveness is not confined to the domain of oxygen alone. When water becomes over-heated, due principally to the increasingly widespread clear-felling of the forest, the health-maintaining pattern of longitudinal vortices changes into transverse ones. These not only undermine and gouge into the riverbanks and embankment works, eventually bursting them, but also create pot-holes in the riverbed itself, bringing even greater disorder to an already chaotic channel-profile.

According to Viktor Schauberger, water subjected to these conditions loses its character, its soul. Like humans of low character, it becomes increasingly violent and aggressive as it casts about hither and thither seeking to vent its anger and restore to itself its former health and stability.

However, due to the senseless malpractice of the clear-felling of forests, we are destroying the very foundation of life. For with the removal of the forest, two very serious things happen:

1). During its flow to the sea, the water warms up prematurely to such an extent that it is warmed right down to the channel-bed. No cool, dense, water-strata remain and the sediment is left lying on the bottom. This blocks the flow, dislocates the channel and results in the inevitable, often catastrophic floods. Yet we still have the effrontery to call these awesome events 'natural disasters', as if Nature herself were responsible. Furthermore, due to the broadening of the channel, even more water is exposed to the Sun's heat, resulting in over-rapid evaporation to the atmosphere. In many cases this overloads the atmosphere with water-vapour, which it is unable to retain in suspension. Deluges follow.

2). With the forest-cover now removed, the ground also begins to heat up to temperatures much higher than normal and natural. Dry soil heats up as much as five times faster than water. This has a two-fold effect:

a). The rejection and repulsion by the warmer soil of any incident rain-water, whose temperature in this case is generally lower. Cold rain will not readily infiltrate into warm soil. This results in rapid surface run-off and no groundwater recharge. The soil dries out.

b). An increase in pathogenic microbial activity, harmful to plant life.

The upshot of all this is more flooding, reduced groundwater quantity and lower groundwater table. One flood therefore begets the next in rapid succession. But since there is no groundwater recharge, the water-balance and natural distribution are completely upset. The remaining trees - the vital retainers of water - die, leaving the land barren and desiccated with the necessary sequel of drought. The less the tree-cover, the more extensive the flooding and the longer the period of drought, of water-lessness, which is synonymous with life-lessness !

Unnatural, quantity-inspired forestry practices, ignorant of Nature's laws, and the over-warming of the soil arising from massive deforestation are the primary causes of the deterioration in water quality, climate and the sinking of the water table. The channelling of water through straight, unnaturally constructed, trapezoid canals, steel pipelines and other misguided systems of river regulation also force the water to move in an unnatural way and accelerate its degeneration and increase its disease-carrying capacity.

> *All around us we see the bridges of life collapsing, those capillaries which create all organic life. This dreadful disintegration has been caused by the mindless and mechanical work of man, who has wrenched the living soul from the Earth's blood - water.*

"The more the engineer endeavours to channel water, of whose spirit and nature he is today still ignorant, by the shortest and straightest route to the sea, the more the flow of water weighs into the bends, the longer its path and the worse the water will become.

"The spreading of the most terrible disease of all, of cancer, is the necessary consequence of such unnatural regulatory works.

"These mistaken activities - our work - must legitimately lead to increasingly widespread unemployment, because our present methods of working, which have a purely mechanical basis, are already destroying not only all of wise Nature's formative processes, but first and foremost the growth of the vegetation itself, which is being destroyed even as it grows.

"The drying up of mountain springs, the change in the whole pattern of motion of the groundwater, and the disturbance in the blood circulation of the organism - Earth - is the direct result of modern forestry practices.

"The pulsebeat of the Earth was factually arrested by the modern timber production industry.

"Every economic death of a people is always preceded by the death of its forests.

"The forest is the habitat of water and as such the habitat of life processes too, whose quality declines as the organic development of the forest is disturbed.

"Ultimately, due to a law which functions with awesome constancy, it will slowly but surely come around to our turn.

"Our accustomed way of thinking in many ways, and perhaps even without exception, is opposed to the true workings of Nature.

"Our work is the embodiment of our will. The spiritual manifestation of this work is its effect. When such work is carried out correctly, it brings happiness, but when carried out incorrectly, it assuredly brings misery."

There is only one solution! Would we live and ensure a sustainable future then we must plant trees for our very lives, but far more importantly, we have a duty to do it for those of our children.

More immediately, however, we must care for the very limited stocks of water still available. This means treating it in the way demonstrated to us by Nature. First and foremost, water should be protected from sunlight and kept in the dark, far removed from all sources of heat, light and atmospheric influences. Ideally it should be placed in opaque, porous containers, which on the one hand cut out all direct light and heat, and on the other, allow the water to breathe, which in common with all other living things, it must do in order to stay alive and healthy. In terms of what we can achieve personally, we should at all times ensure that our storage vessels, tanks, etc., are thoroughly insulated, so that the contained water is maintained at the coolest temperature possible under the prevailing conditions. The materials most suited to this are natural stone, timber (wooden barrels) and terracotta. Perhaps more than any other material, terracotta has been used for this purpose for millennia. Terracotta exhibits a porosity particularly well-suited to purposes of water storage. This is because it enables a very small percentage of the contained water to evaporate via the vessel walls. Evaporation is always associated with cooling (vaporisation, however, with heat) and, according to Walter Schauberger (Viktor's physicist son), if the porosity is correct, then for every 600th part of the contents evaporated, the contents will be cooled by 1°C (1.8°F), thus approaching a temperature of +4°C (+39.2°F).

While the material for the construction of a water-storage vessel has been described above, another important factor is the actual shape of the container itself. Most of the storage containers commonly in use today take the form of cubes, rectangular volumes of one form or another, or cylinders. While these are the shapes most easily and economically produced by today's technology, they do have certain drawbacks in terms of impeding natural water circulation and water suffocation. Due to their rectangular shape and/or right-angled corners, certain stagnant zones are created, conducive to the formation of pathogenic bacteria. Moreover, since the materials used are generally galvanised iron, fibre glass, concrete, etc., i.e. all impervious materials, the contained water is unable to breathe adequately and suffocates as a result. In this debilitated state or as a water-corpse, it is no longer either healthy or health-giving and may require further disinfection.

Should we now make a study of those shapes that Nature chooses to propagate and maintain life, it soon becomes apparent that the cubes and cylinders mentioned above have no place in Nature's scheme of things. Instead, eggs and elongated egg-shapes such as grains and seeds are employed, presumably because Nature in her wisdom has determined that these produce the optimal results. Historically speaking, it is evident that earlier civilisations such as the Egyptians and Greeks, renowned for their logic and constructional ability, were well aware of this, because they stored their grains and liquids (oils, wines, etc.) in terracotta amphorae, sealed with beeswax. All this despite the fact that for all rational, practical purposes, the shape was wholly unsuited to compact and efficient storage in terms of space and ease of handling. It is obvious that the selection of this form over any other was intentional and as the result of certain knowledge of the long-term storage properties of such shapes. In many amphorae that have surfaced in archaeological excavations over the last 100 years or so, grains of wheat have been found that were still viable and even after storage over 2,000 years, grew when planted. This fact alone should suffice to affirm the efficacy of the properties of vessels of such shape.

Taking Viktor Schauberger's exhortation, *Comprehend and copy Nature!* as our guide, we should therefore make use of the shapes that Nature herself selects to contain, guard and maintain life, i.e. eggs and their derivations.

Compared with cubes and cylinders, these shapes have no stagnant zones, no right-angled corners that inhibit flowing movement. By placing our terracotta vessels in shaded areas, exposed to air movement, the evaporative cooling effect will be significantly enhanced and since all natural movement of liquids and gases is triggered by differences in temperature, so too inside the egg-shaped storage vessel, cyclical, spiral, vitalising movement of the water will be induced.

Movement is an expression of energy and energy is synonymous with life. The external evaporation causes cooling of the outer walls and the water in their immediate vicinity. Being cooler and therefore denser, this water becomes specifically heavier and sinks down along the walls towards the bottom at the same time forcing the water there to rise up the centre and move towards the outside walls. Continual repetition of this process results in the constant circulation and cooling of the contents.

Having discussed the above 'ideal' storage vessel and in view of the fact that they are presently not available on the market, it would be a sorry omission, if methods of improving existing installations were not also addressed.

The main factor to be taken into account here is that of exposure to light and heat. Where possible, all above-ground water tanks, whether of galvanised iron, fibre-glass or concrete, should be insulated on all sides and external surfaces through the application of sprayed foam or equivalent thermal barrier to a minimum thickness of 75mm. If not already white or of a light, heat-reflecting colour, then it should be so painted. For in-ground tanks, the top surface only need be insulated and rendered white in colour.

For many people dams or rivers provide the main source of water and certain simple measures can be taken to improve the quality of the water obtained from them.

Providing the surrounding soil is not impervious to water, a hole of suitable dimension, depth and capacity (say 1,000 - 2,000 litres) should be dug about 5 - 10 metres from the banks of the dam or river. If possible the depth should be equal to the depth of the latter. Wells dug next to dams should be situated above the highest water level. If the consistency of the soil is permeable enough, then water will percolate through the intervening soil and into the newly excavated well. Depending on the stability and load-bearing capacity of the soil (a structural engineer should be consulted it there is any doubt), a small concrete, perimeter footing should be placed at a safe and stable distance from the rim of the well. When the concrete has cured and set firmly, then a minimum of 1 course of blocks should be laid to prevent the entry of any surface water. In the case of wells next to rivers, however, it may be necessary to raise the height of the blockwork to just above the average height of flood waters to prevent contamination of the well water during floods.

The well should then be totally enclosed and sealed with a well-insulated timber and sheet-metal roof, or a concrete slab, and provided with an access hatch to service the pump and/or suction pipe and foot-valve. Preferably the pump should be located outside the well-space to avoid any possible oil pollution, etc.

The reason for having the 1,000 - 2,000 litre storage capacity mentioned earlier, is that it may only be possible to pump water intermittently, because the rate of replenishment from the main water source may be fairly slow, depending on the permeability of the soil.

In the event that the soil surrounding a dam or a river is impervious, then it would be necessary to excavate a channel about 600 mm wide between the well and the main water body. The lower part of this should be filled with clean, quartz sand to a depth of about 600 mm and the upper part back-filled with the excavated material and compacted. As the water percolates through either the existing soil or the emplaced sand most suspended matter will be filtered out. Also, because the water comes into the well at the lowest level from the main water source it will be as cool as possible under the prevailing conditions. In this state it is less likely to harbour harmful, pathogenic bacteria, which tend to populate the upper, more highly oxygenated strata of the main water body.

The use of this technique on the author's own property produced an extremely clear, clean, odourless and good tasting water. Despite all outward appearances, however, it is still advisable to have such water tested by the responsible authorities for quality, purity and any possible contaminants.

In terms of its mineral, salt and trace-element content, river-water would generally be far richer than tank-water (rainwater). As for the immature and mature water discussed at the beginning, in most cases it would be necessary to supplement the mineral content of rainwater, if this is the only source of drinking water, in order to prevent the extraction of these from the body of the drinker. Here the suspension of an artificial-fibre sack (rot-proof) containing the dust of crushed basalt or other igneous rock used for road building (commonly known as 'crusher dust') would do much to enhance the composition of the tank water, because it will hungrily absorb those elements it requires to become mature. However, before adding any crusher dust to the water, it would be again advisable to test the resulting change in the quality by analysing the difference between two samples of tank water, one with crusher-dust added and one without as a control. Both samples should then be placed in a cool, dark place and left for at least a week before analysis of the mineral content, bacterial purity, etc. is carried out. This should be done by a suitably qualified specialist.

These suggestions for improving water quality are the result of my personal experience and understanding of Viktor Schauberger's pioneering discoveries and theories.

Viktor Schauberger's great dictum, frequently asserted, was C^2 - *Comprehend & Copy Nature*, for it was only thus that humanity could emerge from its present crisis-stricken condition.

> They call me deranged. The hope is that they are right! It is of no greater or lesser import for yet another fool to wander this Earth. But if I am right and science is wrong, then may the Lord God have mercy on mankind!

Indeed at the Stuttgart University of Technology, West Germany in 1952 these theories were tested under strict scientific and laboratory conditions by Professor Dr. Ing. Franz Pöpel, a hydraulics specialist. These tests showed that, when water is allowed to flow in its naturally ordained manner, it actually generates certain energies, ultimately achieving a condition that could be termed 'negative friction'. Checked and double-checked, this well-documented, but largely unpublicised, pioneering discovery not only vindicated Viktor Schauberger's theories. It also over-turned the hitherto scientifically sacred 'Second Law of Thermodynamics' in which, without further or continuous input of energy, all (closed) systems must degenerate into a condition of total chaos or entropy. These experiments proved that this law, whilst it applies to all mechanical systems, does not apply wholly to living organisms.

As a result of these discoveries, it was arranged that Viktor Schauberger be taken to the United States in 1958, where sums amounting potentially to many millions of dollars could be made available as start-capital for a Los Alamos-like venture to develop Viktor Schauberger's theories of Implosion. He was accompanied by his son, Walter Schauberger, a physicist and mathematician, to assist in the scientific interpretation of his father's theories. Soon after arrival, however, various misunderstandings developed, too complex to elaborate here, whereupon Viktor Schauberger fell silent and refused to participate. After some three months of silence the project was abandoned. Viktor and Walter Schauberger were then permitted to return to Austria, where Viktor died in Linz some five days later on the 25th September 1958, a very disillusioned man.

On their return journey, Viktor asked Walter to translate his theories of Implosion into terms of physics, geometry and mathematics, in such manner that their veracity was irrefutable. Because Viktor Schauberger's concepts broke new ground, this presented some difficulty. There was no adequate scientific terminology to describe them, nor was there any mathematical basis from which the necessary shapes could be precisely defined or constructed. With his own devices and apparatuses, Viktor Schauberger had also encountered problems of construction, which in part affected the optimum functioning of these machines, because the state and sophistication of the technology then available was inadequate and too cumbersome to build them properly and accurately.

The vital development of a new technology, harmonious and conforming to Nature's laws, demands a radical and fundamental change in our way of thinking and to our approach to the interpretation of the established doctrines and facts of physics, chemistry, agriculture, forestry and water management. As a pointer as to how such a new technology should come about, let me quote Viktor Schauberger once more:

> *"How else should it be done then?", was always the immediate question.*
> *The answer is simple:*
> *"Exactly in the opposite way that it is done today!"*

Callum Coats, August 1997

NOTE: All quotations in italics were taken from Viktor Schauberger's writings during the period 1930 - 1933.

The Nature of Water

Viktor Schauberger's overriding passion was the quality of water and the way he perceived that it was being destroyed by contemporary mechanical means. In particular, he raged against what he saw as the devastation of the world's once-sparkling, vibrant great rivers by insensitive hydrologists and river engineers. Driven by the increasing urgency of making these fateful errors known to the public, he wrote *The Nature of Water* and a number of other sections in this book between 1932 and 1933. These were originally published in a two-part book entitled *Our Senseless Toil - the Cause of the World Crisis*. For Viktor, the publication of this, the largest of his individual works, became imperative due to the untimely death in 1931 of Professor Philipp Forchheimer. Forchheimer was a hydrologist of world repute who sponsored the publication of a series of Viktor's treatises on all aspects of water in *Die Wasserwirtschaft*, the Austrian Journal of Hydrology. Without Forchheimer's vital and continuing support, the publication of this comprehensive exposition came to an abrupt end, as is examined at greater length in the main section entitled *Temperature and the Movement of Water* on p.xx - [Editor]

[From *Our Senseless Toil*]

The upholder of the cycles which sustain all Life is *water*[2]. In every drop of water dwells a deity whom indeed we all serve. There also dwells Life, the soul of the primal substance - water - whose boundaries and banks are the capillaries that guide it, and in which it circulates. Every pulse beat arising through the interaction of will and resistance is indicative of creative work and urges us to care for those vessels, those primary and most vital structures, in which throbs the product of a dualistic power - Life.

Every waterway is an artery of this Life, an artery that creates its own pathways and bridges as it advances, so as to diffuse its dawning life-force through the Earth and elevate itself to great heights, to become shining,

---

[2] The basic teaching of the Ionic natural philosopher Thales (625-545 BC), *'Water is the source of all Life'*, embodies a profound understanding and is of great importance. It should in no way be construed as idle speculation. As a Greek, he had intuition, which according to Goethe is "a revelation emanating from the inner self". Intuition is spiritual seeing, not an insight gained through experience or rational thought. According to Spinoza, it is the highest form of perception because the pure principles of Nature alone remain active and the categorising propensity (*compart*-mentality) of the human mind does not come into action. It is therefore unable to perceive the world as a *multiplex unitas* and necessarily creates a highly incomplete picture. - VS.

beautiful and free. Standing at the highest level of evolution, and above all being blessed with mind and reason, humanity constantly does the most idiotic thing imaginable by trying to regulate these waterways by means of their banks - by influencing the flow mechanically, instead of taking into account the fact that water is itself a living entity.

The assumption behind this absurd practice is that the riverbank shapes the watercourse, whereas the riverbank is actually the secondary effect and *water the primary*. To regulate water by means of the riverbank is truly to fight cause with effect. It should be as inconceivable to a thinking engineer to reinforce the crumbling bank of a watercourse with rammed piles and brush-wood bales, or to smear over cracks with cement, as it would be for a doctor to patch up ruptured capillaries with needle and thread. Astonishingly, though, it still happens! The condition of all our waterways demonstrates just where these measures have led.

In not one single case has the desired object been attained - namely the achievement of a normal channel-profile. On the contrary, all such river regulation has provoked further damage which far outweighs any local or short-term advantages. Large rivers such as the Danube, Rhine, Tagliamento, Etsch, Garonne and Mississippi bear witness to the failure of such complicated and costly river regulations. Quite apart from the tremendous damage caused in the lower reaches by their strictly mechanical regulation, these rivers are stripped of their most valuable assets, their great physical qualities.

The present dirty grey, muddy brew known as the Blue Danube, upon whose bed river-gold once gleamed, and the Rhine, the symbol of German identity, where Rhinegold flashed in bygone days, are tragic testimonials to these perverse practices. The mythical `Gold of the Nibelungs` originated in the golden glow given off by pebbles as they rubbed against each other while rolling along the riverbed at night - for when there is a decrease in water temperature, the tractive force[3] increases, causing the stones to move. If two pebbles are rubbed together under water, a golden glow appears.[4] This yellowish-red fiery glow used to be mistaken for the flashing of gold, thought to be lying on the bottom of the river. Today this `river-gold` lies heaped up in huge mounds of gravel, shifted hither and thither by the force of the sluggish and murky water-masses[5] flowing above them. They no

---

[3]*Tractive force*: This refers to the force described hydraulically as 'shear force' - the force that acts to 'shear off' or to dredge and dislodge sediment. In German the term for shear force is 'Schubkraft', meaning 'to push, to shove' as well as 'to shear', whereas Viktor Schauberger uses the word 'Schleppkraft'. The verb 'schleppen' means to drag, draw or pull. Viktor Schauberger's choice of 'Schleppkraft' here is quite specific, since in his view the movement of sediment is due to the sucking action of fast flowing, dense cold water downstream, rather than to the mechanical impact of the water coming from upstream. In view of this subtle change in emphasis, in lieu of the hydraulically correct term 'shear force', the term 'tractive force' will be used. This dynamic is similar to the effect of wind on roofs, where a roof is blown off not by force from the windward side, but rather by the sucking effect of vortices created on the leeward side.- Ed.

longer imbue the water with *energy* and *soul,* as once they did. Instead they assist in ousting the soul-less body - water - from its badly-regulated course.

Our clear, cold mountain streams have become wild torrents. Full of the vigour of youth, these lively streams used to be surrounded by burgeoning vegetation and consorted with every blade of grass as long as man did not interfere. Today they can no longer be confined even with metre-thick concrete walls. Wherever we look we see the dreadful disintegration of the very

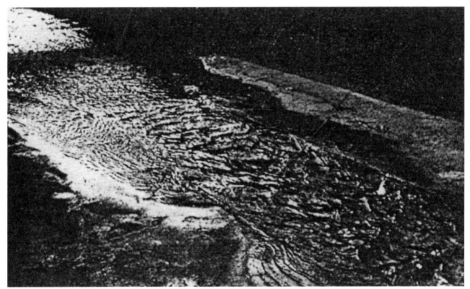

Fig 1: The confluence of the Tepl and the hot-spring at Karlsbad (now Karlovy Vary in the Czech Republic). The inflow of warm water provokes the formation of transverse blocking currents in the water masses. (Negative temperature gradient, i.e. the temperature of the water masses deviates from the anomaly point of +4°C [+39.2°F].) Note the barren river bank, destruction of the riverbed and bank. Water-masses which flow under a negative temperature gradient destroy the channel.

bridges of Life, the capillaries and the bodies they have created, caused by mindless mechanical human acts. These actions have robbed the Earth's blood - water - of its soul. It is therefore inevitable that the larger and more expensive these regulatory structures become, the greater will be the ensuing damage. In the lower reaches of the Danube almost a million hectares of valuable farmland have been lost due to the regulation of the upper reaches. Similar conditions apply to all other rivers.

[4]This is a phenomenon akin to triboluminescence, which relates to the light given off by crystalline rocks under friction or violent pressure. It is attributed to the energy emitted by the electrons contained in the rocks as they return from a pressure-induced excited state to their rest orbits. - Ed.

[5]*Water masses*: this expression refers not only to the body of water generally, but also to the various swirling volumes and filaments of water of different temperatures, densities and energetic content whose values are prescribed by the inner dynamics of the water. - Ed.

Even today the river engineer fails to understand the true nature and purpose of water. The harder he tries to conduct it by the shortest and straightest route to the sea, the more it will tend to form bends, the longer will be its path and the worse its quality. The flow of water down a natural gradient obeys a sublime, inner Law whose power our hydraulic experts are quite unable to comprehend. In the absence of this inner conformity with law, all flowing water ought to accelerate faster and faster until it ultimately transfers to a vaporous state. Science maintains that water is braked by internal and external friction, though it is well known that friction is associated with the generation of heat. However, it can be shown that the temperature of fast-flowing water *decreases*, which leads to an increase in tractive force and internal friction. This simple observation invalidates certain essential propositions in the complex of current hydro-mechanical theories.

Where then is the real secret of steadiness in the flow of draining water-masses? The force that brakes the flow of water down a gradient is a resistance which acts against the force of gravity, a circulation of energy operating in the opposite direction to the current. This is also true of all metabolic processes and gives water its character and thus its soul. Contemporary systems of river regulation inhibit this vital function. The logical outcome of this is the loss of water's inner braking power. The water becomes soulless, without character and therefore aggressive.

Fig. 2: The confluence of the Tepl and the Eger. The Tepl, previously warmed by an affluent hot-spring, cools off in the lower reaches. (Positive temperature gradient, i.e. the temperature of the water masses approaches the anomaly point of +4°C [+39.2°F].) Note the fertile river bank, the narrowing of the channel cross-section and the straight flow of water. Water masses which flow under a positive temperature gradient build up the riverbank

The Cancerous Decay of Organisms
[From *Our Senseless Toil*]

The more extensive regulatory works become, destroying water's naturally-ordained inner functions, the greater the ensuing danger to the riverbanks and the surrounding area. Now *characterless*, the water breaks its bonds. Having become unstable, it seeks to regain its soul with one last supreme effort. The water-masses abandon their proper course and countless water-borne energy-bodies are dropped by the exhausted water. Disoriented, it now turns on these organisms and robs them of their life-force. Deprived of their souls, their sources of energy, they begin to rot. Bacteria develop and the Earth's arteries are suffused with cancerous decay.

Sinking into the ground, this diseased water now contaminates groundwater. As it rises through the capillaries of the soil and vegetation, this very Blood of the Earth carries the embryo of this fearful disease with it up into the widest variety of plants. This leads to the qualitative degeneration of vegetation, principally in the internal decay of forest trees. As a further consequence, it leads to a regression in the quality of everything in which water circulates. Ultimately in accordance with a law which operates with awesome constancy, it will slowly but surely come around to our turn. The spreading of the most terrible of all diseases - cancer - is the inevitable consequence of these unnatural systems of regulation. It goes without saying that specialists in other fields also have a hand in this work of destruction.

The Substance - Water
[From *Our Senseless Toil*]

"The revelation of the secret of water will put an end to all manner of speculation or expediency and their excrescences, to which belong war, hatred, impatience and discord of every kind. The thorough study of water therefore signifies the end of monopolies, the end of all domination in the truest sense of the word and the start of a socialism arising from the development of individualism in its most perfect form."
Viktor Schauberger. 1939 - *Implosion* Magazine, No.6, p.29.

By taking the right paths we are led back to Nature and hence to the source of life, to healthy water. The higher up such water springs forth from Mother Earth the healthier it is. Borne up by inner energies it emerges only when *ripe* - when it has achieved its proper physical composition and when it must leave the Earth. The absorbed air content of such water consists of about 96% gaseous, physically dissolved carbones.[6] As a result the *psyche* or the *character* of water can be described as being of very high calibre.

There are some springs which exhibit such a high content of carbonic acid (this expression is in any case incorrect) that when small animals inhale the vapour from it in the surrounding atmosphere they die almost instantly - the Dog Spring in Naples being an example.[7] This water is also dangerous for people who suck it into their mouths directly from the spring, and inhale its rising gases at the same time. Mountain folk call such springs 'poisonous water'. Today springs can still be found which people avoid and which are fenced off to prevent access by grazing cattle. According to folklore these springs are inhabited by 'Waterworm' which, if swallowed, irrevocably cause death within a few days. If a metal container is filled with such water and placed in the open air, the water heats up inordinately quickly, displaying a slight effervescence at the surface. Incidentally this phenomenon also sometimes happens when wells are bored. Whenever these events occur, the exposed water quickly subsides and the well is soon dry. When such water is exposed to the air, the emergence of an abundance of bacterial life is soon observed. The warmer the water becomes, the less complex and the more primitive the bacteria. If warmed-up rainwater is poured into this water, a few drops of oil are added and the whole container sealed, the contents of the container will very soon explode.

What has happened here? The negative atmosphere, the *psyche* contained in the high-grade springwater oxidises. It interacts with warm, heavily-oxygenated and consequently predominantly positively-charged air. When this interactive expansion encounters an obstacle and when a high-grade carbone is present - for example, in the form of oil - it shatters the container.

If such water is drunk quickly when the body is hot, the same phenomenon occurs in the body of the drinker. The affected person feels a stabbing pain in the lungs and dies within a few days. Alpine farmers describe this rapid sickening as the `vanishing lung disease` (galloping consumption). If such cases are less frequent today than in earlier times, this is only because such high-grade water is now rarely found.

By means of the above interactions, energies will be either freed or bound. The defining factors connected with this encompass the diverse composition of the atmosphere and the varying effect of light, both of which are con-

---

[6]*Carbone:* In contrast to the normal use and definition of 'carbon', Viktor Schauberger grouped all the known elements and their compounds, with the exception of oxygen and hydrogen, under the general classification of 'Mother Substances', which he described with the word 'Kohle-stoffe', normally spelt 'Kohlenstoffe' and meaning carbon. Apart from the above definition the hyphen also signifies a higher aspect of carbon, both physically and energetically or immaterially. The additional 'e' in the English word is therefore intended to redefine and enlarge the scope of the usual term 'carbon' in accordance with Viktor's concepts. On occasion *carbone* will be represented by the term C^e to differentiate it from the normal term for carbon - C. - Ed.

[7]This emerges in a subterranean cavern, producing a layer of pure carbon dioxide (also known as `choke damp`) immediately above it, which suffocates any straying dogs. Human beings, however, being taller, can breathe the air above this carbon dioxide `sea` and survive without lasting injury. - Ed.

ditional on the season and the height of the Sun. The longer the water is exposed to the influence of light and the more it comes into contact with the air during flow or through mechanical motion (stirring action), the more it will relinquish its former geospheric characteristics,[8] absorb those of the atmosphere and become warm, stale and insipid.

The more immature (juvenile) the water emerging from seepage springs or otherwise extracted from the ground, and the smaller the difference between contrasting magnitudes originally present, the weaker will be the interactions. The more inferior the quality of the products of this energy-exchange, the less complex are the micro-organisms which evolve. This necessarily results in the mental and physical degeneration of all those organisms that use this low-quality water in order to function. If the vitally important oxidising processes can no longer occur in an appropriate high-grade form, it is unreasonable to expect to find high-grade properties and processes continuing in water that is no longer able to maintain its inner ripeness or has lost its previously mature characteristics. It should therefore come as no surprise if, in such water, a variety of more primitive life-forms come into existence which ultimately pose a threat even to human life.

Whereas an initial supply of oxygen was necessary for the emergence and development of these organisms, an excessive concentration of oxygen or an over-supply of a lower-grade oxygen would tend to be fatal to them. The same sort of thing happens to us. If we wish to visit the sphere of oxygen, the stratosphere, we must take with us oxygen of the same composition normally encountered in our own sphere. The same holds true for the supply of fresh water on ocean voyages. If excessive quantities of oxygen are injected into water then, in the long run, such water can support neither bacterium nor human being. Since the bacterium has no other way to breathe it must die immediately - whereas the human being, who at least still has a chance to gulp down some healthy air, perishes only in the course of time.

The body's metabolic processes are dependent on a specific composition of basic elements - the carbone and oxygen groups contained in water. Similarly, the development of qualitatively high-grade vegetation is dependent upon there being a particular ratio between the quantities and qualities of these substances in the basic formative substance - water. These quantities and qualities generate a particular *internal temperature* appropriate to each organism in which they are taken up (whether by breathing, through the consumption of food or the direct supply of water) as a result of the interactions occurring during these reciprocal oxidising processes.

---

[8] The actual expression here is 'Earthsphere'. Where Viktor Schauberger uses the word 'sphere' in this context, this denotes the collective embodiment of all the physical, material and energetic characteristics inhering in any given sphere, be it the hydrosphere, geosphere, biosphere, atmosphere, etc. - Ed.

A particular inner temperature produces a certain physical form which in turn generates the special kind of immaterial energy we encounter as *character*. Hence the old saying *"mens sana in corpore sano"* (a healthy mind in a healthy body). If the composition of basic substances should in any way be altered, not only must the metabolic basis for further growth of the body change but so must its spiritual and intellectual growth and further development.

Briefly summarised: healthy air, healthy food and healthy water produce not only a healthy body but also good character traits.

Concerning Processes of Ur-Creation, Evolution and Metabolism
[From *Mensch und Technik* Sec. 6.0, vol.2, 1993]

The following section is taken from a special edition of the German quarterly periodical, *Mensch und Technik - naturgemaß* (Humanity & Technology - in accordance with Nature), which is devoted entirely to the then recently discovered transcript of a notebook compiled in 1941 by a Swiss, Arnold Hohl. This not only recorded details of his visits to Viktor Schauberger in 1936 and 1937, but included verbatim accounts of Schauberger's contemporary writings, letters, notes and comments, recorded by Hohl verbatim. Funded by private subscriptions, *Mensch und Technik* was first published by a group of scientists, calling themselves the *Gruppe der Neuen* (Group of New Thinkers). Their aim was to explore Viktor Schauberger's theories and to interpret them scientifically. Early articles came from a number of contributors including Viktor Schauberger himself (posthumously) and his son Walter. - [Editor]

Joking aside, making water is not at all a simple matter, for it requires intuitive understanding. Every drop of water is virgin territory and a source of unlimited power. According to science, horsepower by the thousand exists in every gram of water.[9] Yet dare we take the plunge to extract it? This process is of extreme importance, completely natural and therefore simple too, for water indeed embodies the unity out of which the immeasurable multiplicity subsequently arises.

Water-eggs are present in both metal and mineral. They should be produced and allowed to incubate from these substances. A little heat, a little cold, a little light and a little darkness are quite enough so to inflame the passions of metal and mineral alike that they produce a chemically pure

---

[9] According to Dipl. Ing. Walter Schauberger, Viktor's physicist son, every gram of any substance - 1 cubic centimetre of water for example - contains a stored energy equal to 25,000,000 kilowatt hours. - Ed.

water. Once this stage has been reached, the whole begins to grow spontaneously. All it depends on is the proper preparation.

The phenomena and processes of transmutation which arise from differences in energy potential and produce water-eggs, permeate and animate the remaining substances. Being in a less organised state, the latter would never experience their transformation into an organically higher form of life if they did not actually force the juvenile or virgin water to carry out purposeful work before their ascension. It is these substances that lift the water up and not the water that raises them. Precisely the reverse takes place in the high atmosphere, where it is the juvenile water in this case that forces the very buoyant substances to conduct the water *ur*-procreated in the atmosphere gently back to Earth. In other words, the juvenile water ensures that the heavenly substances imbued with mechanical levitative energies are continuously returned to Earth in order to assist other substances to reach higher regions and levels of evolvement. Thus in Nature mechanical pressure and physical suction are always in a condition of rhythmical alternation, which is why there is no state of rest in this world, for the relation between pressure and suction is always in the ratio of 2 : 3. This concept is so simple that it requires a certain finesse not to reveal too much, because it would be absolutely no blessing to humanity, if with its present attitudes it was suddenly given control over these elemental forces.

The most important precondition for the production of water is the "angle", for through it the feelings find their expression, such as happens with laughter. Without pulling the angles of the mouth, laughing would be impossible. "$^\circ\cup^\circ$" or "$^\circ\cap^\circ$". This angle is a "oneness" in a monolinear plane, infinite and very similar to "∞".

This infinite plane is what we describe as "time". All we have to do is to create organic loops[10] in order to produce water-eggs and from them the water itself, which on its part then creates the spaces for nutritive material. These water-eggs are even sometimes visible to the naked eye, because we are here concerned with products of the anomaly point (for water = +4°C), which are incubated with cold instead of heat. Once such a process of incubation[11] has begun, then the heat is dissipated in the water, because all processes of growth require heat. This is why rivers that cool themselves as they flow increase their tractive force and carry the "naughty" stones, which lie so heavily in the stomachs of river regulators, far out into the seas and create new land. In the opposite case, they create water-deserts.

Protein Formation: This can appear in many different shapes and sizes. Protein is in the earth in a solid state, in water in a liquid state and in air in a

---

[10] Organic loops may perhaps take the form of figures-of-eight or be akin to the Moebius strip, which has one face and one edge - Ed.

[11] In-*cube*-ation = turn into 3 dimensions. - Ed.

gaseous state. In all cases these are manifestations of concentration or *ur*-combination. We thus come face to face with the processes of *ur*-procreation.

Solid, liquid and gaseous are three spacial states. If by means of the appropriate angle, we concentrate the ethereal elements or ethericities[12] into a suitably dosed mixture of the three groups of substances, then we are presented with the natural formation of protein. In order that the process should produce the desired effect, it is very important that we pay strict attention to the different properties of the tensions existing between water and air. Therefore we have to organise pressure and suction in the proper directionally alternating rhythm, which from a physical point of view are actually forms of heat and cold. This is enough to produce any desired quantity of protein artificially, but in the way that it happens in Nature. In this way proteinic substances can be produced in their solid, liquid or gaseous ur-form. Consequently there are also water-eggs, air-eggs and earth-eggs.

- Air-eggs produce water.
- Water-eggs produce earth.
- Earth-eggs produce energy-eggs (dynagens).

With these energy-eggs we are thus presented with the possibility of producing a flowing movement of energy. Space-energy propulsion, which is actually within the constitution of such eggs, is therefore already within our grasp.

These energy-eggs are themselves nourished by the substances of the air; that is, a vacuum is formed when we set this process in motion. This vacuum, however, is totally different from the "nothing" that science seeks to describe as a zero. Those entities or magnitudes hitherto considered to be virtual or marginal will become the actual causes from which can be derived not only any desired equivalent factor, but also significant elemental forces as well, once the dosing and organisation of these dangerous opposites is understood and they are brought into a rhythmically oscillating motion. Since pressure and suction are constantly in an opposing mechanical and physical state of unstable equilibrium, then all we have to do is to organise the rhythm. We are thus presented with ceaseless motion that constantly increases. This is a mystical phenomenon, today quite inconceivable, but which nevertheless is what the natural motion we can see and detect all around us.

---

[12]*Ethericities:* This term refers to those supranormal, near non-dimensional, energetic, bio-electric, bio-magnetic, catalytic, high-frequency, vibratory, super-potent entities of quasi-material, quasi-etheric nature belonging to the 4th and 5th dimensions of being. As such they can be further categorised as fructigens, qualigens and dynagens, which respectively represent those subtle energies, whose function is the enhancement of fructification (fructigens), the generation of quality (qualigens) and the amplification of immaterial energy (dynagens). According to their function or location these may be male or female in nature. - Ed.

- If oxygen is infused into water indirectly, then the water will become colder.

- If carbon-dioxide is absorbed by water indirectly, then the water will become colder.

For every 1°C that water is cooled, the volume of its contained gases is reduced by 1/273rd. Cooling water transforms its gases into volume-less substances. This juvenile energy dissolves metallic and mineral solids, through which heat becomes bound, which in turn absorbs the heavy oxygen. The carbone content of winter-water is minimal and hence there is little absorption of oxygen. On the other hand, winter-water also becomes lighter and absorbs atmospheric gases, when for lack of suitable minerals, it cannot obtain carbonic acid from the earth.

Springwater high in carbones does not even freeze at -32°C. Such water takes up oxygen in summer and becomes colder under high external temperatures. Under certain circumstances carbon-dioxide is heavier than both air and water, whereas oxygen under certain conditions is lighter than air and heavier than water. The underlying cause of this remarkable variation in weight is the degree of excitation of the gas. This is why the weight of water constantly changes, and it is important to differentiate between specific and absolute variations in weight. The higher the water rises in the spring-conduit, the heavier the minerals it precipitates. The more it unburdens itself, the denser it becomes and the more easily it rises.

Water is energetically discharged by iron grilles and in turbines. That is to say, it loses its dynagens, becomes unipolar and in the lower reaches replaces its lost dynagens from the surrounding soil. This is why such water destroys its channel. The water becomes an animalistic magnet as it were, and in eviscerating the soil of its dynagens it also rips soil-particles away. In winter the water attracts aeriform substances, because it is element-hungry. In this high state of tension, however, it is unable to metabolise them and therefore becomes lighter.

On the other hand the atmosphere can also attract water if the water is terrestrially overcharged. When vapour rises under conditions of extreme cold, however, this is conditioned by the water's inner atomic charge.

- Water changes its boiling and freezing points according to its inner state of tension.

- Water is no lifeless mass, but the blood of the Earth, which comes into being and disappears through energetic interactions.

- Trees synthesise water. Hence no forest - no water.

The latter is the product of radiant emissions, which are instrumental in forming the transmuting medium at the root-tip. This protoplasm, erroneously called a suction cap, can be seen with the naked eye on every root-tip of a forest tree. With illumination it collapses and disappears. The formation of sap is also the result of radiation and thus has nothing to do with any mechanical suctional or pressural activity. The same is true for our blood circulation and the upwelling of springwater.

If water is devoid of inner atomic energies it is unable to rise. This also applies to the tractive force and the deposition of sediment. Under-energised blood leads to accretions in, or sclerosis of, the conducting vessels. The supply of energy-water alleviates or removes this affliction. All kidney diseases are to be attributed to de-energising phenomena. For this reason kidney-stones are immediately expelled if 'noble' water is drunk.

High-Frequency Water
[From *Implosion Magazine No. 24.*]

Implosion **is a quarterly magazine, originally published from about 1958 by Aloys Kokaly. Kokaly's association with Viktor Schauberger began in 1940, when he manufactured certain components in Germany for Viktor's so-called 'flying machines'. These, according to Viktor's physicist son, Walter, were smuggled into Austria where they could neither be obtained nor manufactured. This fairly close association continued right through the war and up to Viktor Schauberger's death in 1958. Shortly thereafter Kokaly founde** *Verein zur Förderung der Biotechnik e.V.* **(Association for the Advancement of Biotechnology) specifically for the research and evaluation of Viktor's theories and discoveries. In an effort to reach an even wider audience, conferences and seminars were held annually at the Association's headquarters in Neviges, Wuppertal.** *Implosion* **was published in order to provide a platform for Viktor and Walter Schauberger's various writings, of which Kokaly had many originals. -** [Editor]

Throughout the world hunger will be constantly on the increase and is the result of the improper conduction and conservation of water. Wherever people are hungry, a state of strife must also exist. For this reason it is necessary to investigate the underlying causes to which peoples of ancient civilisations owed their thriving vegetation, their several harvests a year, in short, their whole culture - a culture that ultimately perished, however, due to over-cultivation.

In Nature two forces prevail:
- **gravitation** - for expulsion and purging,
- **levitation** - for upward impulsion and synthesis.

In the struggle between the *Ur*-Feminine - the formative life-principle - and the fertilising, husband-like *Ur*-Masculine, *cycloid motion* plays a decisive role in determining whether a rise or fall in living standards results.

The state of equilibrium between expulsive and impulsive forces always remains constant.[13] If the weight of the organisation of basic elements is reduced due to the disturbance of cycloid motion, then it is a sign that the build-up of quality has also been disturbed - a build-up that is only possible with the aid of cycloid, oscillating motion. If cycloid motion is fostered in draining water, then the growth of *qualigen* is increased. *Higher-potency substances or potentialities* are built up. The ensuing increase in the growth of raw materials then acts as a counterweight to the intensified buoyant or levitative force, so that in all cases the labile state of equilibrium is maintained. Between these two mutually self-compensating interactive forces - the up-building and down-grading - like a finger on the scales of Fate, which indicates the present state of well-being like a barometer, stands the current living standard.

A decline in living standards is indicative of the neglect of cycloid motion. Conversely, the standard of living rises immediately if the propagation of this form of motion, to which everything owes its material existence, is reinstated. Without cycloid motion there can be no development of quality matter and hence no purging of those substances too deficient or unsuitable for the purposes of higher evolution.

Viewed naturalesquely, all growth and vegetation, indeed all manifestations of life, are inferior energy-concentrates of raw materials or basic elements. This becomes clearer if water, the `first-born`, or the first form of crystallising-out into matter, is viewed as a non-uniformly heavy carrier-substance. Seminal or fertilising matter, also known as oxygen, has a different type of charge and for this reason also has a different atomic weight to the stocks of oppositely-potentiated fecund matter (hydrogen and carbones). Only through cycloid motion can the latter be transformed and built up into free potentiated entities. Only then is the consumption (binding) of the seminal substances (oxygenes) by the real formative elements possible. The product of these transformative processes is high-frequency water.

If the seminal elements, which react neither to pressure nor tension, bind the germinal substances, then heat is created. The water loses its original buoyant energies, becomes stale and inevitably degenerates. The enormous loss of formative energies can perhaps be estimated if one considers that when a large river is warmed by 0.1°C per cubic metre, formative energies

---

[13]This constancy is founded on the reciprocal relation between the two magnitudes of expulsion and impulsion. - Ed.

are lost which correspond to a power equal to 42,700 kgm.[14] If this water, which due to the absence of cycloid motion has become warm and stale, is caused to move cycloidally on a trial basis, then it approaches the anomaly point of +4°C regardless of the Sun's influence. At this temperature elements of carbonic acid bind elements of oxygen, and therefore the loss of fertilising substances is to the gain of the stocks of fecund matter now coming to life. Water thus conducted becomes fresh, increases its carrying capacity and tractive force and surrenders its surplus formative energies to the surrounding soil. There the groundwater is immediately recharged, with the result that heat is extracted from the ground. This results in the build-up of the germination zone - the boundary zone between the positively-potentiated atmosphere and the negatively-potentiated geosphere, which must be located within the vegetation root-zone - if these fecund elements are to develop further and be able to surrender their surplus formative energies to root-protoplasms. It is the same process that occurs in the human body, where the normal build-up of blood can only take place at the organic anomaly point of between +36°C (96.8°F) and +37°C (98.6°F). If the blood temperature rises above +40°C (104°F), then blood decomposition sets in. Similarly, when the temperature of riverwater rises significantly above its anomaly point, then all possible growth of vegetation ceases and it becomes dessicated and shrivelled. This also happens with the lowering of the ground temperature in winter, when the build-up of sap, the real accumulator of formative energies, is de-activated.

The pulsation of the blood improves its quality and facilitates its circulation. No circular motion is implied by the circulation cycle, but rather a gyration along a circular path. In its spiralling inward-and-outward oscillation, it has an action similar to a clock spring. Through this lively motion it is constantly shaken up, jostled and jolted, which enables the constant and continuous uptake of nutrients, water and air to be maintained. The more rhythmical the impulse-giving movement (as in certain physical exercises), the healthier the organism becomes. Due to this highly favourable movement, the inner pulsating dynamics of the blood and, in a broader sense, the force of its spiritual potency is maintained.

Therefore, when a river does its physical jerks and constantly sways from right to left and from left to right again, a strong renewal of water ensues, leading to the consumption of irradiated heat-producing matter, which then becomes bound, due to the vigorous growth of water and formative energies. The oxygen becomes cool and inactive, whereas the fecund elements become free and highly active. It is only in this way that the proper process of synthesis can take its normal course.

[14] *kgm*: kilogram-metre [mechanical] = 1 kilowatt hour [kWh] = 36,700 kgm. In this case therefore 1.163 kWh. - Ed.

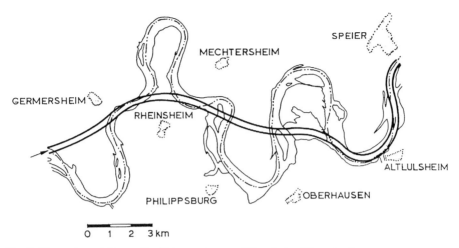

Fig. 3: Channel regulation on the Rhine upstream of Mannheim (19 century)

If a river is straightened out, as occurs with modern river regulation, to shorten the course by truncating the meanders (*see fig. 3* of the Rhine) and steepening the geological gradient, then variously-weighted particles of sediment are precipitated laterally and vertically, according to their specific weight. The river deepens its bed on the outer curves and raises its bed on the inner curves by sediment deposition on the warmer riverbank. As a result of this regulation the surface-draining bloodstream of the organism - Earth - is ejected from its developmental path both longitudinally and transversely, and in the process is *thoroughly finished off.*

However, if we can picture a naturalesque river regulation, which incorporates raised banks on the outer bends of the channel similar to those of a bob-sleigh run or a railway, then it becomes clear why such a water-course supplies the whole environment with high-grade, formative energies (*see figs. 4 & 5*). The channel for the water must be so arranged that, due to its momentum, the descending water over-falls or over-reaches itself and accelerates towards the centre - the flow-axis - in increasingly smaller spiralling whorls. This is where the centrally-transported, fertilising substances are concentrated, because they do not react to centripetence (*see fig. 6*). It is thus equally clear why female, fecund elements are able to ensnare male seminal elements and ultimately consume them entirely. In which case everything around begins to flourish and burst into blossom. Conversely, the fertility of the soil sinks all the more when, due to an unnatural form of flow, the *straightened* water, now no longer able to do its normal physical exercises, is *mercilessly killed off.*

High-frequency water is therefore water in which organic processes of synthesis can take place naturesquely. In such water fecund elements, i.e.

NATURALESQUE RIVER REGULATION (patented)
(Cheap, quickly installed, indestructible)
(Gill-combs of fish are delicate, flutter and are in a state of equipoise)
(between pressure and suction)

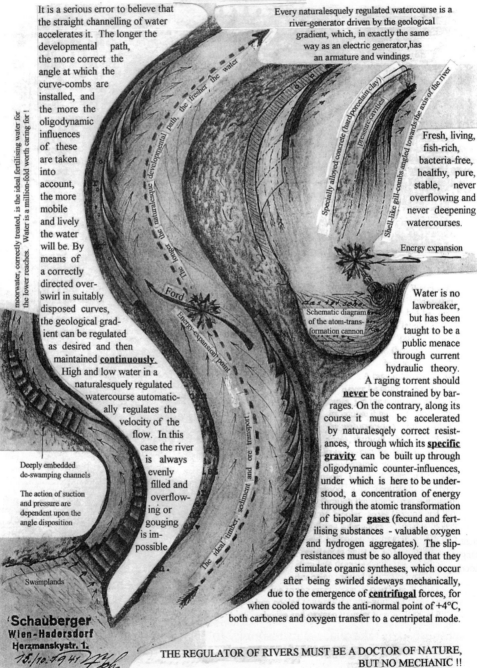

It is a serious error to believe that the straight channelling of water accelerates it. The longer the developmental path, the more correct the angle at which the curve-combs are installed, and the more the oligodynamic influences of these are taken into account, the more mobile and lively the water will be. By means of a correctly directed over-swirl in suitably disposed curves, the geological gradient can be regulated as desired and then maintained **continuously**. High and low water in a naturalesquely regulated watercourse automatically regulates the velocity of the flow. In this case the river is always evenly filled and overflowing or gouging is impossible.

moorwater, correctly treated, is the ideal fertilising water for the lower reaches. Water is a million-fold worth caring for!

Deeply embedded de-swamping channels

The action of suction and pressure are dependent upon the angle disposition

Swamplands

'Schauberger
Wien-Hadersdorf
Herzmanskystr. 1.

Every naturalesquely regulated watercourse is a river-generator driven by the geological gradient, which, in exactly the same way as an electric generator, has an armature and windings.

Specially alloyed concrete (hard porcelain clay)
pressure cavities
Shell-like gill-combs angled towards the axis of the river

Fresh, living, fish-rich, bacteria-free, healthy, pure, stable, never overflowing and never deepening watercourses.

Energy expansion

the longer the naturalesque developmental path, the fresher the water

Ford
energy expansion point
The ideal timber sediment and ore transport

Schematic diagram of the atom-transformation cannon

Water is no lawbreaker, but has been taught to be a public menace through current hydraulic theory. A raging torrent should **never** be constrained by barrages. On the contrary, along its course it must be accelerated by naturalesqely correct resistances, through which its **specific gravity** can be built up through oligodynamic counter-influences, under which is here to be understood, a concentration of energy through the atomic transformation of bipolar **gases** (fecund and fertilising substances - valuable oxygen and hydrogen aggregates). The slip-resistances must be so alloyed that they stimulate organic syntheses, which occur after being swirled sideways mechanically, due to the emergence of **centrifugal** forces, for when cooled towards the anti-normal point of +4°C, both carbones and oxygen transfer to a centripetal mode.

THE REGULATOR OF RIVERS MUST BE A DOCTOR OF NATURE, BUT NO MECHANIC !!

Driven naturally by the geological gradient, **every** naturalesqely regulated river is a **river-generator** or the **progenitor** of products of organic syntheses, which directly serve the growth of the vegetation through the agency of the charge-accumulator, the groundwater.

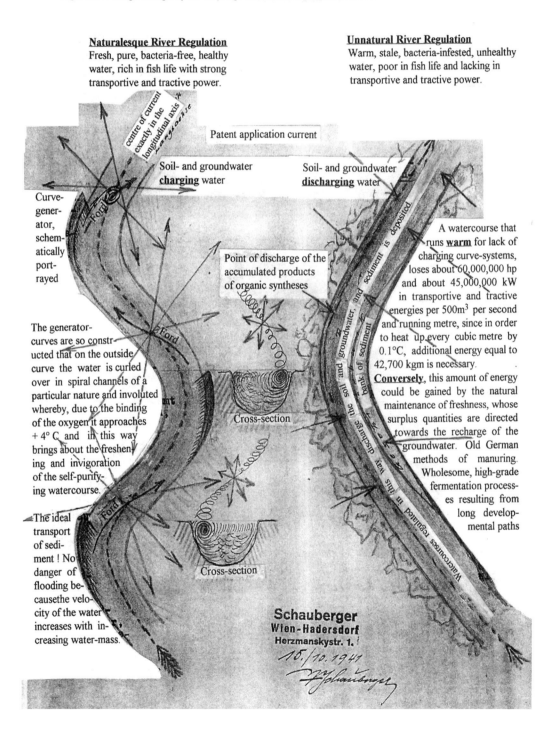

Naturalesque River Regulation
Fresh, pure, bacteria-free, healthy water, rich in fish life with strong transportive and tractive power.

Unnatural River Regulation
Warm, stale, bacteria-infested, unhealthy water, poor in fish life and lacking in transportive and tractive power.

Patent application current

Soil- and groundwater **charging** water

Soil- and groundwater **discharging** water

Curve-generator, schematically portrayed

Point of discharge of the accumulated products of organic syntheses

A watercourse that runs **warm** for lack of charging curve-systems, loses about 60,000,000 hp and about 45,000,000 kW in transportive and tractive energies per 500m^3 per second and running metre, since in order to heat up every cubic metre by 0.1°C, additional energy equal to 42,700 kgm is necessary.

The generator-curves are so constructed that on the outside curve the water is curled over in spiral channels of a particular nature and involuted whereby, due to the binding of the oxygen it approaches + 4° C and in this way brings about the freshening and invigoration of the self-purifying watercourse.

Cross-section

Conversely, this amount of energy could be gained by the natural maintenance of freshness, whose surplus quantities are directed towards the recharge of the groundwater. Old German methods of manuring. Wholesome, high-grade fermentation processes resulting from long developmental paths

The ideal transport of sediment! No danger of flooding because the velocity of the water increases with increasing water-mass.

Cross-section

Schauberger
Wien-Hadersdorf
Herzmanskystr. 1.

15./10.1941

NATURALESQUE RIVER REGULATION

As the generator of rivers, the curve within the curve is the sexual organ of water.

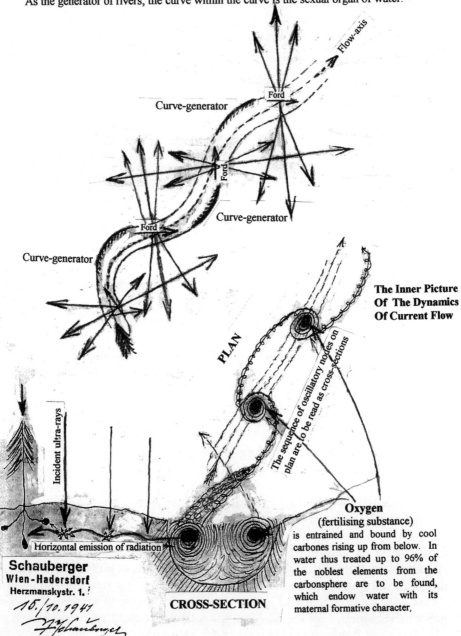

Oxygen (fertilising substance) is entrained and bound by cool carbones rising up from below. In water thus treated up to 96% of the noblest elements from the carbonsphere are to be found, which endow water with its maternal formative character.

correctly-evolved growth factors, dominate, whose ever-increasingly high grade waste products *solidify in equally increased measure*, due to the strong influences of heat and light. As inferior physical organisations of raw materials or basic elements, they are condemned to vegetate, the ecological sequel of which is an over-supply of vegetation or food.

Therefore let water do its physical jerks! Grant it its rights to swing about to its heart's content! Then there will be a super-abundance of food, and eternal peace will come about of its own accord.

Viktor Schauberger, Vienna, September 1943.

The Natural Reconversion of Seawater into Fresh Water
[From *Implosion* Magazine No. 101.]

This discovery is founded on the realisation that from a biological point of view there are two different kinds of mass motion and hence equally differing forms of atomic energy - animating (propagating and upwardly evolving, or multiplying and ennobling) as well as de-animating (retrogressive, repulsive and atavistic). Both types of motion operate simultaneously on a common developmental axis. The form of atomic energy that ultimately prevails depends on the type of impulse that this rhythmical motive process imparts. The art of regulating this process was known and practised by the High Priests of ancient cultures, who knew how to order the eternal metamorphic flow *(Panta Rhei)* in such a way that the demand for food and raw materials of a constantly increasing population could be satisfied. The untroubled continued growth at the time is to be ascribed to this regulatory skill. Lost thousands of years ago, this art was rediscovered through long years spent observing the motionless stance of the stationary trout amidst fast-flowing water and is now discussed:

Growth-enhancing atomic energies arise when the media of earth, water and air (the carriers of stocks of bi-polar basic elements) are made to move in a planetary manner. In this case it is possible to combine basic elements of opposite potential - solidified solar energy, chemically described as oxygen - by means of the intrinsic valencies and properties of the sediments contained in water and air, if, under the total exclusion of light, heat and air, these media are caused to inwind and at the same time both stocks of basic elements are atomised and highly excited by means of suitable catalysts. In this regard, however, care must be taken to ensure that retrogressive atomic forces only become free and active to the extent required to expel the waste matter left behind after the raw materials have been reduced. As faecal matter, they must be removed if unwelcome accretions or decomposition of the above residues are not to occur in the organism in question.

Aware only of centrifugal mass motion, it is today's school-aristocrats (academics) who have disrupted this wonderful reduction-purification process. Following natural law, it is therefore inevitable that in all four areas of industry (forestry, agriculture, water resources and energy) a retrogressive development will proceed all the more rapidly the more these scholastics, trained in technical colleges and universities to become demagogues, achieve their supposed successes in mass motion and excitation. For all motion, they use devices, machines and conveyors whose back-to-front construction and fabrication with the wrong alloys fosters those atomic energies that inhibit evolutionary syntheses. Having unwittingly provoked a decline across the full spectrum of human development, it is therefore the academics, who are to be regarded as the true inhibitors of evolution or as the long-sought instigators of cancer.

It goes without saying that today's leaders of industry will find this assertion both outrageous and offensive. If one considers, however, that as a direct result of the above mistakes in mass motion, thousands of millions of their innocent fellow human beings have been robbed of their most fundamental right to existence; that as a result of the lack of high-quality substances they have become cancer-prone and are decaying alive, ultimately to die prematurely with unspeakable suffering - then no word is trenchant enough publicly to denounce this travesty of motion.

"Life-force" can only be maintained through the supply of additional animating energies. These become free and active when, like ingested food, the earthly residues of life have undergone a kind of digestive process. For this the original, planetary, movement of mass is required, which in the organism of the Earth creates a counter-climate to that of the atmosphere, namely the state of indifference or the anomaly - for water +4°C, which has sovereign power over the decisive "to be or not to be" interaction between basic elements.

The High Priests of ancient cultures were able to control this climate so skilfully that in all forms of life and growth they were able to maintain the so-called anomaly - temperaturelessness and feverlessness - the state of health as constant as possible. Varying from individual to individual, the respective life-form bears the hallmarks of its own particularity. Accordingly, whenever an individual becomes ill it is a sign that it is no longer able to maintain this condition of indifference. In other words, it has become overcooled, overheated, has consumed genetically impaired food or feverish water, or has somehow breathed in contaminated air. In short, it has ingested over-acidified concentrates of energetic matter.

If still water is excessively bombarded by the unbraked (unfiltered) rays of the Sun, for example, or if flowing water is forced to move academically (hydraulically), then it will become over-acidic and imbued with an unhealthy potential the organism is unable to withstand. It sickens as a result, spoils and, slowly proceeding further on the path of degeneration, dies.

Just as no-one has hitherto succeeded in producing artificial seawater in which sea fish can survive, so is it also true that, apart from me, nobody has managed to create a nutritive liquid, correctly constituted physiologically, by artificial means, otherwise known as healthy springwater. This is because nobody has so far taken note of the fact that so-called 'mountain springwater' can only be created through the agency of the Earth's planetary motion. Together with certain other co-active factors, this type of motion conditions the particular climate in which naturalesquely atomised solar energies - chemically termed oxygen - as well as atomised sediments or their released intrinsic attributes (atomic energies in a nascent state) can be bound (emulsified). Emulsion means to intermix the oppositely-potentiated ethericities of basic elements - so-called emanations - so intimately that, as happens with any other process of genesis, a third something emerges. In other words, a child is born, which is either a male or a female entity, or a hermaphrodite, depending on the type of fertilisation. The hermaphrodite, however, can subsequently be transformed into a predominantly male or female life-form through the appropriate influences.

Mountain springwater comes into being, or the water-child, the physically first-born blood of the Earth, is *ur*-procreated when geospheric and atmospheric rays intimately interbreed under diffuse influences of light and heat with the presence of catalytic secondary radiation. This water decomposes, dies and sinks away again if the spring is over-illuminated and over-warmed, exposed to direct sunlight - in other words, over-acidified.[15] Such over-acidification can also occur when initially healthy (about +4°C) water is subjected to excessive solar irradiation on its course, or if it is wrongly (hydraulically) regulated in improperly profiled and/or conducted in unnaturally alloyed pipes and channels, or is centrifuged in any way by means of iron (steel) motive devices (pressure turbines, pelton wheels, pressure pumps, etc.), i.e. if it is handled academically, namely in accordance with conventional scientific doctrine.

The blood of the Earth is predominantly maternal, or formative in nature. For this reason it must in-whorl about its ideal axis in exactly the same way as Mother-Earth. Day by day she moves around the *ur*-fertilising Sun, which constantly alters the angle of incidence of its rays so that no over-fertilisation and therefore no over-acidification can ever occur.

The academic (hydraulicist and dynamicist) make this pressure and heat sensitive blood of the Earth move centrifugally almost exclusively, namely

---

[15]Here the reader is referred to footnote 25 concerning the German word for oxygen, *Sauerstoff*, in the light of which the over-acidification mentioned here could also be interpreted as over-oxygenation, or the result of it. - Ed.

from the centre of the axis towards the walls. In short, in an out-winding manner. Because of this, hitherto unknown ultra-red rays with a character akin to x-rays and additional heating effects of a reactive nature come into being, which make the oxygen present in all water aggressive; its counterpart on the other hand, which is released through the corrasion of sediment, is made passive (inactive). As a result this triggers a reversed interaction between the basic elements; the end-product of which is the de-animating or life-removing form of atomic energy.

This type of emanation radiates in all directions, overcoming all resistance and penetrating right through to the negatively charged cell-protoplasm or sediment-nucleus. This it heats up, splits apart, thereby turning the previously healthy life-cell into a new epicentre of decay, whose emission of increased and intensified radioactive (life-destroying) rays (see the discovery of Professor Hahn)[16] contaminate all other forms of life and growth. That is to say: they become infected with an atomic embryo of corruption and are turned into the carriers of genetic disease (charged with cancer).

An apparently harmless mechanical, physical or psychic excess pressure is enough to inaugurate localised processes of decomposition - the well-known cancerous tumours - that we can observe particularly clearly in over-illuminated and overheated shade-demanding timbers. The enlargement of the annual rings, the increasingly spongy consistency of the wood, i.e. so-called *"light-induced growth"*, has hitherto been taught in all colleges and universities by the academically trained forester as an *achievement* of forestry science.

In doing this he unwittingly enforced the destruction of quality substances in today's monocultivated young forest and made it thoroughly sick with cancer. The next generation had already lost its power to reproduce and improve in quality, and bore infertile seeds. And thus, like water that has become genetically diseased due to so many retrograde influences - it too is irretrievably lost.

Whatever takes place within the sap system of the plants, also has similar consequences in other nutritive beds such as blood and groundwater. This is why the present alarming increase in avalanche disasters is also the biological and therefore the foregone sequel to the structural disintegration brought about by academic errors in motion and influences. The old-growth stands of timber, more firmly rooted in their natural habitat, have fallen victim to irresponsible over-felling. Riddled with white and red rot, these wooden cancer-impregnated (spongiform) structures can no longer withstand the pressures of wind and snow, nor counter the shearing force of the

---

[16]This refers to the German physicist Otto Hahn who, in collaboration with the Austrian nuclear physicist Lise Meitner, discovered the radioactive element *protactinium* in 1917 and who with his fellow German physicist Fritz Strassman demonstrated the nuclear fission of uranium when bombarded with neutrons. - Ed.

snow, because the over-illuminated and over-heated ground, devoid of trees and other vegetation, has also become cancerous right down to the root-zone. Through the resulting change to its state of potential and being unaffected by further changes in temperature, the ground lost its bio-magnetic, attractive power in relation to the positively over-charged snow, and thus, in accordance with the principle - like repels like, the snow began to slip even on gentle slopes and in this way avalanches happened where they had never before in living memory.

In both cases we are actually concerned with the forces of atomic pressure and suction, which also explain the original processes of sap and blood circulation for the first time. It is necessary to understand these processes precisely and to know how to regulate them correctly if the natural metabolic cycles are to be maintained, so that atomic energies of a predominantly growth-promoting and upwardly evolving, i.e. animating, nature are to be brought forth indirectly through naturalesque processes of emulsion (*ur*-procreation).

Accordingly we are here concerned with hitherto undetectable and unmeasurable interactions between emanation-essences, i.e. with the principle of countercurrents, whose bipolar mixture of rays possesses an intrinsic velocity of such a magnitude that it has first to be decelerated naturalesquely in order to be able to verify its existence with appropriate measuring instruments, or before it can be recognised by means of various light-effects in vacuum tubes as an animating or a de-animating form of atomic energy.

It would take too long at this point to provide more detailed explanations, since in principle academics unknowingly generate the same annihilating energies with back-to-front methods of mass motion that the nuclear physicist, as a supposed means of maintaining peace, intentionally generates mechanically for the radical extermination of all life. The former energies act as 'insidious poison' and are the true cause not only of the general condition of physical retrogression and economic decline in society, but also of the ever more widespread incidence of cancer. For this reason the whole of humanity is beyond all hope of recovery, unless the academic errors in mass motion can be eliminated.

In order to reconvert over-acidified fresh water or seawater into genetically healthy drinking and usable water naturalesquely, this most valuable national asset must be conducted in such a way that it can synthesise recreative and refreshing atomic energies once more.

These energies can then bind and emulsify excess oxygen. The product of this relatively simple regenerative process is not only genetically healthy drinking and utilisable water, which can once more develop and qualitatively improve - multiply and ennoble - itself, but it should also be noted in passing that it is also Nature's cure-all, which heals and prevents the regressive course of cancer. In this so-called fresh (sweet) water even sea fish are able to exist, because it secretes within itself those animating atomic energies without which there can be neither life, nor autonomous freedom of movement on Earth, on and in water and in the air.

Just how people and above all our children can be protected from a ghastly future and how the Earth may be transformed once more into a true paradise has been demonstrated through the granting of pioneer patents and through instructions supported by university expert opinion.[17]

These will signal the end of present methods of generating electricity, which of course are also based on mechanical ways of moving water, because naturalesquely operating hydroelectric turbines have not only almost doubled the output, but also serve as the ideal machines for river regulation. This will be discussed later, however, when the occasion arises (see Patent No. 117749 in Appendix).

With the collapse of academic demagoguery, the social mass-privation and with it the retrograde development of the spiritually castrated (robbed of quality substances) humanity will also be at an end, for without the contribution of animating atomic energies, individual mental vigour is quite unthinkable.

Fire Under Water
Schauberger Report, written by Werner Zimmermann.
[From *TAU*, August, vol 148, pp.9-11.]

Founded in 1924 in Zurich, Switzerland, *Tau* (German for *Dew*) was a periodical, which described itself as a monthly magazine for *"spiritual perception and personal growth, awareness and action"*. **Sustained by private subscription, it was compiled and published by Professor Werner Zimmermann and was probably one of the earliest so-called 'new age' publications. It concerned itself with all aspects of culture in its highest sense, with articles ranging from raw food to the teachings of Krishnamurti. Naturally aligned to Viktor Schauberger's insights into the inner workings of Nature, it provided an ideal vehicle for the wider dissemination of his ideas. Schauberger and Zimmermann worked together through some of Viktor's most difficult years (1935-1936), when Viktor fought long and hard battles against the authorities to save both the Rhine and the Danube from total ruin and rehabilitate them through proper, naturalesque regulation. Regardless of the personalities and institutions involved and threats to his own person, Zimmermann fearlessly reported these events, blow by blow, to his readers. - [Editor]**

We have by no means solved all the enigmas of water. Viktor Schauberger has several original concepts which are perhaps worth verifying. He really should spare us his Faustian fables: that rolling pebbles can emit sparks

[17] viz. the Stuttgart investigation by Professor Franz Pöpel. See the companion volume *The Energy Revolution* - Ed.

under water; that water grows; and that the stones at the bottom of a riverbed are the river's bread.[18] It is hard to take such a man seriously!

Several scientists, quite independently of each other, have made such comments to me about Schauberger's explanations. I told Schauberger about these instances. He laughed his hearty laugh, doubtless sensed my over-riding desire to see something of these wonders, and asked me: *"Sparks under water? Have you never seen them yet? Then I'll show you right away. Come!"* He took two pebbles from a pigeonhole, poured water into a bucket, took it over to a dark corner, and lo, all at once as he rubbed the stones together in the water, they gave off sparks just as they would have had they been rubbed together in the air. Cold light!!

Now for me that was a fairy-tale experience, and all those to whom I have since demonstrated this play of sparks in water have had similar reactions. It really stood a lot of things on their head, and at last gave me access to many insights which Schauberger had presented to us.

Schauberger: *"Once in Yugoslavia I was really made to look a fool. I wanted to show the sparks to several unbelieving people and collected two pebbles from the river, but in no way could I achieve the slightest glimmer of light. It thus became clear to me that in their long way along the riverbed, these stones had lost their inner energy. You should therefore select pebbles from the mountains or the upper reaches of a river. But best of all, break a pebble in two and use both halves."*

I will now try to express what these underwater sparks have made more understandable to me. For this I cannot avail myself of customary scientific terminology because I am not in command of it and furthermore, because it must lose much of its present validity. Words have not been developed for these new concepts and facts. This has made it much more difficult for Schauberger to make his theories understandable, and thus he always referred to the facts of the matter as he saw them whenever he spoke of them.[19]

In the case of the underwater sparking pebble-stones however, we have the advantage that we are here concerned with a fact which can be verified relatively easily by anyone without any previous theoretical experience. It is all the more astonishing that there still must be professors and engineers concerned with hydraulics and river regulation to whom it has not even

---

[18]Zimmerman is referring to Schauberger's writings in *Tau* 137, p11, *Tau* 144, p29, *Tau* 146, p29, etc. In *Tau* 137, pages 10-11, Schauberger wrote: *"If two completely identical pebbles, or two identical, flawlessly grown, fine timbers are rubbed together under water, they give off a clearly detectable fiery glow - a gleam of light. Why do two completely identical bodies produce fire when rubbed together? What is it that is burning? Why isn't this fire extinguished by the surrounding water? Why does this fiery glow become even more intense when the water approaches its characteristic, anomalous temperature of +4°C?"* - see footnote 3 on triboluminescence. - Ed.

[19]As for example in *Tau* 146 where he wrote on "Dancing Sinkers and Stones" in an article entitled "The Ox", and in *Tau* 147 on "Excommunication and Little Wandering Springs" - see Vol.II of *Eco-Technology: Nature as Teacher* – Ed.

once occurred to put this simple experiment to the test - since for them even the very thought of it is impossible. It is in the nature of revolutionaries and laymen to be creative - such is not the case with the typical run-of-the-mill expert and scientist. To see with new eyes and create new things, one must boldly step beyond tried and trusted pathways into the unknown.

Schauberger has asserted much that seems fantastic from my personal insights and practical experience. At first just about all I could rely on was my finer inner feelings. When I was convinced of the reliability of his statements, I gave him my willing support. Every subsequent practical verification, however, has shown that in the main Schauberger stands firmly rooted in true reality.

Schauberger sees two polar forces at work in everything, which we can denote as male and female, insemination and giving birth, spiritual and material, active and passive, positive and negative, *Yang* and *Yin*, Heaven and Earth. (Roundness and point, egg and orifice are respectively the polar forms of creation.) Further poles are: Sun and Earth, Sun and Moon, gold and silver, copper and tin. Both are of equal value, and in their interplay these forces create new combinations and forms.

Water too is such a creation, whose principal basic elements we know as hydrogen H and oxygen O. Oxygen is active in all combustion, in all disintegration and new construction, as a male force, whereas hydrogen belongs to the opposite aspect (female), which Schauberger now and then incorporates in the concept of carbone or *mother-substance*.

Fire is the result of a union between a father-substance, O, and a mother-substance, C, H, etc. In a certain sense even water as hydroxide, H_2O can be construed as a product of combustion (not that we can thereby comprehend it in its living totality). So just what do sparks under water imply?

1. Through rubbing stones, heat arises not only in the air but also in water;

2. Thus in a stone, forces or substances are released which are to be described as female;

3. In water there must be free oxygen (male), similar to the air, otherwise no flame could be created in the water;

4. If an oxide can actually be created in water by a ray of flame, similar to a marriage between stone (mother-substance) and oxygen (father-substance), then we could also conceive that in such a manner, perhaps in a less forceful process, the oxide - water - can also be created.

I have no illusions that with this simple observation I have actually *proven* anything. It suddenly dawned on me what sort of process it might be that Schauberger observed and described as the growth of water (*Tau* 147, p22 -

"Excommunication and Little Wandering Springs", which permitted him to consider sediment as the bread, the provisions, for the water's journey - and why he also warned so insistently against dredging.

Now I can read with much greater understanding Schauberger's reply to Kobelt (*Tau* 146, p29) in relation to the cooling of the Rhine. We must learn entirely anew. With hands empty and ready to receive, we must climb up into the mountains towards the dawn. Only then shall we begin to understand the effects set in motion by double-spiral pipes. Yet such organic machines, which are now being experimentally built, will still appear as wonders to us. Every marvel, however, is none other than the result of conformities with a law which we still do not understand, and are thus not able to set in motion.

Notes on the Secret of Water
[From *Implosion* Magazine No. 46.]

Today it is known that water is not just `water`. There is also *heavy* water, for example, which has very specific properties. Also water is known in its various states of aggregation as liquid, vapour and ice, as an inorganic substance (so it is believed), which moves from cloud to ocean and back again. Armed with this knowledge it is usually considered that the most essential aspects of water have been exhaustively explained. In relation to the study of water movement:

It was by studying water movement over the years that I discovered the enormous difference between the higher and lower grade means of acceleration of mass;

Wise Nature uses what I call the *original* motion - the pressure-free, heatless and thus *resistance-free* form of motion - by up to 90%;

X-rays belong to the developmentally-dangerous forms of atomic energy. In the resistance represented by the fabric of the body they trigger off warm light-effects which, with reactive force, enlarge cell tissue (loosen the structure). For this reason alone X-rays are character- or quality-decomposing;

That even the smallest effects of pressure on water trigger the release of dangerous or harmful X-ray-like radiation is proven by the following experiment:

As many as 2,000-4,000 volts are produced and directly measurable on an electrometer if a drop of water is merely allowed to fall a few centimetres due to its own self-weight, and due to the pressure exerted on it by passing through a needle-jet. If the jet pressure is raised to between 2 and 3 atmospheres, then the measuring device already indicates up to 15,000 volts.

If this downward flow is collected on a metal strip, then 2,200 volts per droplet can be accumulated. This charge is then conducted by an insulated wire to a vacuum tube, which lights up with a strongly pulsating, dark red glow. If the charge is allowed to accumulate in a Leyden jar and is conducted from this into petrol (gasoline), it ignites this with a jet of flame, whose magnitude increases as the charge in the jar rises.

This demonstrates that this form of energy is predominantly *electrical*, therefore water-decomposing (or blood- or sap-decomposing). It involves the discharge of analysing energies in ray form. Whether the water is pressed against a wall, or a wall (such as a ploughshare) is pressed against the water (soil moisture), is immaterial, for in both cases this developmentally-harmful, decomposive current appears *if* the needle-jets or the pressurising surfaces are made of *iron*. In the case of a ploughshare it is dangerous because it decomposes the blood of the Earth - water.

However, if a jet or pressure-imparting surface (ploughshare) made out of *copper* is used, which is so shaped that a drop in pressure occurs with acceleration - then an apparently identical current capacity is produced quantitatively, but on closer examination this is predominantly *magnetic*. It *draws together* the surrounding material through which it passes and therefore has a structure-condensing or quantity-reducing function.

The Production of Fuels
[From *Mensch und Technik*, Spec.ed, relating to Rivers, (Vol.2, 1993, sec.8) Author's notes from 7 July 1939] See introductory comments to *Concerning Processes of Ur-Creation, Evolution & Metabolism*, p.xx

In this chapter we shall discuss how to replace the energies derived from oil, coal and water with high-grade fuels, which as waste-matter arising from processes of potentiation and depotentiation, can be obtained from water and air. Water and air are tripartite systems, which are mutually related under conditions of contra-directional potential.[20] For this reason certain differences should be taken into account when decomposing these two substances. In addition, it is necessary to have a general idea of the origin of these groups of substances, what they are and what purpose they serve.

Viewed biologically, water is an accumulator as well as a transformer; that is, under certain preconditions certain substances can be introduced into or extracted from water. Water is thus an organ, which is remotely controlled by the cathode and anode systems of the Sun and Moon. Accordingly it can

---

[20]This refers to the different, but reciprocally related directions in which the respective energies of water and air are propagated. - Ed.

carry out anodic or cathodic functions depending on the given circumstances.[21]

In all processes of solution and combination, water is governed by various influences. The most important, which lead to transformative processes within water itself are a) light and heat, and b) darkness and cold. The metabolic processes in water, therefore, are engendered by the alternation of night and day or the life-rhythm of the Earth. Conversely, metabolic processes trigger dynamic phenomena and thus we are presented with a practical *'perpetuum mobile'*, which is easily copied, once one comes to understand this ceaseless process of Nature's just a little.

The first stage in reconstituting water is to free it of all solids. Inasmuch as it is practicable, most of the bipolar gaseous elements should also be removed. Purging the water of matter is most easily achieved by depotentiating it, namely the removal of the substances that cohere within the water-body itself. Up to now little attention has been paid to these organic, magnetic forces. Few people suspect that water is an accumulator of high-potency, vital energies or electrozoic[22] essences. These dynagens, which are absorbed by our bodies when we drink water, for example, are organic plus and *minus* impulses, i.e. vital life-renewers. For this reason it matters a great deal what kind of water we drink or use for cooking. Every laundry knows that *aqua destillata* (distilled water) and rainwater, for example, have markedly different properties in their ability to attract dirt.

Whether drinking water deposits energies in our bodies or extracts rejuvenating substances is merely a question of the temperament or the will that the water possesses at the time of drinking. For example, if the water-structure's cohering energies are removed from the carrier substance, then all the solids fall to the bottom and most of the gaseous elements escape upwards. Depending on whether it is more strongly or more weakly discharged, the residual material is a more or less polarised substance, or manifests itself as a deficiency, which has the will to recharge itself with substances taken from anywhere. This is why this polar substance acts like a magnet.

If we can imagine this depotentiation process in its extreme form, then what is left as residual matter is an ur-substance of such deep valency[23] that the most gigantic metabolic processes can be unleashed, because quite cata-

---

[21] An anode is an electrode carrying a positive charge to which negatively charged *anions*$^-$, also electrons, are attracted. Similarly, a cathode is an electrode carrying a negative charge, to which positively charged *cations*+ migrate. In this sense therefore, the condition, movement and energetic character of water, its resultant interaction with its surroundings, respond to the fluctuating influences of the oppositely charged Sun (cathode) and Moon (anode).

Elsewhere, in *Healing Water for Human, Beast and Soil* - p.xx, Schauberger refers to *cathode-water* and *anode-water* which respectively carry high percentages of carbone-energies and oxygen-energies. - Ed.

[22] *Electrozoic essences* can also be interpreted as animalistic or organismic essences. - Ed.

[23] highly polarised, pregnant status - Ed.

strophic transformative activity arises through its reconstitution. This can of course be put to practical use, providing an appropriate machine has been built in which the contrasting elements can interact in a useful way. Apart from the practical advantages, this also brings other benefits. For example, natural catastrophes could be nipped in the bud, if, at an early stage in their evolution, their innate energies were used to power machines.

The Difference between Energising Substances and Fuels
[From *Mensch und Technik,* Spec.ed, relating to Rivers, (Vol.2, 1993, sec.8) Author's notes from 7 July 1939] See introductory comments to
Concerning Processes of Ur-Creation, Evolution & Metabolism, p.xx.

If the quickening substances, which in Nature are intended for growth and increase, are misguidedly exploited for powering machines, then it is no wonder that coffins are becoming more common than cradles. The best evidence of this malpractice are the present methods of streamlining river drainage by straightening the channel, by bank rectification or the construction of dams in a manner that drags in by heels precisely those catastrophes that river engineers are trying to avoid.

The form of the channel, its profile and the curves in which the water flows is the negative or mirror-image impression of the forces active in the water. Any flow of water, which weighs too heavily into the bends, forms meanders or actually eats into the riverbank or destroys it, is sick. If its natural flow is brutally constrained by the riverbank, then what was originally "good-natured" water will become increasingly diseased, malicious and dangerous until it dies and seems to disappear. It then returns with punitive vengeance to confound those who robbed it of its health, and who denied its very existence.

Where does this strange power in the water come from? Who are the gods who can confer benefits just as easily as ruthlessly punish?

In order to understand these contra-directional forces, we must learn to observe with precision and to understand the kaleidoscope of interactive colours and effects, and indeed the *ur*-language of water. It is a living world of whose might and power the experts have no conception. The confinement of god-like entities or life-energies with rammed piles, concrete walls, etc. could only be undertaken by those who have no idea of a life-force that can move mountains and transform waste matter into wholesome food.

The Quantitative and Qualitative Detoriation of Water

[From *Our Senseless Toil*]

At a time when the treatment of water by chemical means was being generally hailed as a great breakthrough in public health and safety, Viktor Schauberger's voice was one of the few cautionary voices being raised against its potentially harmful consequences. - [Editor]

The Deterioration of Water

For about a decade the groundwater has been sinking so fast in many areas that the number of years before people will be forced to abandon their upland villages and homes can be counted on one's fingers. This is either because their vital water supplies will have ceased to exist or will be obtainable, if at all, only at great cost.

With the sinking of the groundwater table, springs peter out, streams dry up and the soil, which ought to provide our daily bread, dies of thirst. In other places where water rises out of the Earth again, rivers break their banks and turn the countryside into swamps. In addition to this alarming quantitative shift in the distribution of water in, on and above the Earth, an even greater danger threatens: the *qualitative deterioration of increasingly-scarce residual sources of water.* This will render not just drinking water, but even domestic water, directly harmful to health.

Just how far the latter danger has already advanced is clearly evident in a Press article concerning an investigation of the water in London's reservoirs and swimming pools. It appeared in the *Daily Mail* on 23rd August 1933. These investigations established clear proof of the presence of over a million bacteria per cm^3 in the water of public swimming baths - places where thousands of people seek recreation, yet who thereby expose themselves to serious contagious diseases. If this danger already exists in constantly-monitored facilities, how much great it might be where such controls are absent. Apart from these revelations this investigation produced yet another surprise: it was established that where attempts had been made to remove this danger through chlorination, bathers experienced serious inflammation of the eyes and mucous membranes of the nose.

The Sterilisation of Water

One of the most difficult tasks in the treatment and preparation of drinking water involves the sterilisation of *surface water and* immature (juvenile) groundwater which endanger health and are unsuitable for drinking purposes. As a rule this water is taken from rivers, lakes and reservoirs or, where these sources are unavailable, is pumped up from deep wells and rendered (theoretically) drinkable with the use of chemical additives. All those forced to live in cities are well acquainted with the bad taste of mechanically-filtered water, water which is *contaminated by micro-organic* matter, artificially polluted by chlorine, irradiation or other sterilising agents and disinfected by chemical compounds and other ingredients. What are not known, however, are the consequences ensuing from this.

While the dreadful repercussions accruing from the continual consumption of sterilised drinking water may not be clear to water-supply engineers, doctors cannot claim to be unaware of the causes of the sickness appearing everywhere. Their responsibility is all the greater since they are the ones who are supposed to keep the organic formation of the body and its various stages of development under constant observation and study. In view of the fact that contemporary doctors must also acquire certain preliminary technical knowledge and an understanding of various basic chemical and physical principles before beginning their medical studies (which to a large extent will actually rob them of their connection with reality), practising physicians should at least understand what effect the continuous consumption of sterilised water will have on the human body and whether the continued use of this method of sterilisation should still be permitted.

Those doctors who dedicate their whole lives to cancer research and are adequately supported financially in their endeavours should first of all ask themselves the question: how does such bacterial life evolve in the human body or in any other organically-constituted body? It is not sufficient merely to record the existing facts and to try to eliminate the unwelcome life-forms that already exist. The very fact that the development of bacteria is enhanced in water - water left standing for long periods, flowing slowly in the sunlight or in badly-enclosed, open wells - must point to certain correlations which urgently need to be researched, in order to put an end to the danger of disease associated with them. If this path has not yet been trodden, it is because our practising physicians themselves have already lost touch with Nature.

In the final analysis, all attempts to purify drinking water are directed towards creating conditions unfavourable to the bacterial life that evolves in it under certain conditions, in the hope of eradicating it. If the water has been rendered `hygienically impeccable` in such manner, then, as a rule, one is entirely satisfied with it and believes that enough has been done. Quite apart from any other associated hazards - for example, residual micro-organic matter unpurged by present systems of sterilisation - it never enters anyone's head that certain material energies will also be denied to people who regularly consume sterilised water, sterilised milk or other sterilised foods. This deficiency will lead to a decrease in their mental, physical and sexual potency and will inevitably increase the health risks to their weakened bodies. After a lengthy time of constantly consuming water treated in this way, the blood will be systematically destroyed. This enfeeblement leaves the door wide open to the entry of disease.

The Consequences of Chlorination of Water

In the increasingly difficult matter of supplying drinking and domestic water to cities and housing estates, scant consideration is given to the water's content of suspended solids. In addition, its *internal physical processes* and *character* are also completely neglected. As a rule, chlorination is deemed satisfactory to obtain clear, pure and germ-free water.

There is hardly a city where water is not disinfected or sterilised through the addition of chlorine, compounds of silver or irradiation with quartz lamps. In all these processes oxygen *in statu nascendi*, or an allotropic form of common oxygen, is produced which will kill off all living organisms. If water thus treated is drunk continually then the very same processes that we wish to achieve through water sterilisation must also take place in our bodies. Frightful consequences can ensue from the constant consumption of such water. When sterilisation only is taken into account, there arise the various forms of the disease we collectively call *cancer*. In 1920, 2,400 people died of cancer in Vienna; in 1926, 3,700 fatal cases of cancer were recorded; in 1931, 4,900 fell victim to this terrible illness. From these figures the progressive spread of this disease is clearly evident.

This dreadful scourge which, despite all the efforts and skills of our medical research institutes, can neither be accurately recognised for what it is nor effectively controlled, and whose spread affects more and more victims, is primarily an after-effect of unhealthy or badly-conducted water. This not only contributes to the chemical make-up of our food and the constitution of our blood, but also determines the quality of the composition of the atmosphere directly surrounding the organisms inside the body. Relevant statistical data clearly reveal that cancer is most prevalent in those districts where no good, high-quality springwater is available. Even in those places where the springwater is still good and healthy, it will deteriorate as a result of being transported in pipelines sometimes hundreds of kilometres long.[24] The emerging pattern of the spread of cancer can be measured against the length of the pipes in which domestic and drinking water flows to its point of use.

This assertion will be immediately countered by the statement that the water has undergone all conceivable tests and its day-to-day content of dissolved and absorbed matter is accurately monitored. If we continue to drink sterilised water only, we must also accept the ensuing consequences. If we do not wish to suffer a slow death in mind and body, we must search and strive for other ways to cast the Devil out of today's drinking water, but not with the Devil himself!

The Consequences of Contemporary Water-Purification Processes

The quality and quantity of water's oxygen content is substantially altered under present systems of aerobic water purification which take place under the influence of light. This immediately results in disturbances to the metabolism. As a further consequence, it results in aggregations of oxygen which the water in the body, already over-saturated with oxygen, cannot assimilate. As a result of the additional internal pressures thus created the first symptoms of disease manifest themselves as swellings and tumours. In the case of trees, these become clearly visible in shade-demanding timbers exposed to direct sunlight, in warm, strongly-oxygenated soils.

The presence of excess oxygen in the enlarged cells leads to production of high levels of acid[25] and subsequently to inflammation. This inflammation in turn engenders even higher temperatures - fever - causing the oxygen to become increasingly aggressive and eventually to compensate for the lack of any other carbones by combining with the substances of the tissue itself. This results in the emergence of inferior and less complex microbes which, under suitable preconditions, begin their vital activity. In the absence of other food they make a regular feast of the macro-organism - the body. The disease-causing organism is therefore the indirect product of incorrect metabolic interactions. Science describes this is as cancer.

The only means of defence presently available to science is the knife or radiation. Were our doctors to understand why cancerous tumours really

---

[24] Only recently another 427 km long water main has been laid in southern California to supply Los Angeles with water. - VS.

[25] The German word for oxygen - *Sauerstoff* - and its root - *sauer* - throw some interesting light on this effect. Directly translated *Sauerstoff* means 'sour stuff'. Further derived from this is the German word for 'acid', namely *Säure*, which seems to affirm a direct connection between oxygen and the formation of acids- Ed.

begin to flourish once the body is opened up, or were they able to comprehend the underlying causes of combustion phenomena (inflammation), they would no longer use these methods.

It is a remarkable fact that distilled water greedily absorbs gaseous substances from the surrounding atmosphere, so that it soon takes on the smell of the substances surrounding it. Because such sterile water extracts gaseous carbones from its environs, medicine has also made use of it to purge human blood. The consumption of such water can only bring about a short-lived improvement in the general condition. In the most favourable cases it merely acts as a stimulant but, in the final analysis, such water can only act destructively on the organism, since it ultimately removes carbones from it. In this case these are not excess waste products, but the most vitally important formative substances.

The beneficial effect of completely sterile water, therefore, can only be of brief duration, since the surrounding medium - the body - is divested of its most highly essential substances. This then serves to create the breeding ground for new micro-organisms. If attempts are made to sterilise water by chlorination alone, some of the oxygen will still be retained after the disinfecting activity of the aggressive oxygen has ceased. When this encounters the requisite particles of carbone it triggers the formation of microbial life in no uncertain measure. The carbones in water should be viewed as negative electrons and the oxygenes as positive electrons which, under the influence of temperature and in conformity with natural law, are mutually opposed in inverse proportion.

If we take in good food, good air and healthy, *mature* water, highly complex bacteria are formed which consume the low-grade life-forms that may eventually develop. If on the other hand we ingest inferior raw materials, whether in low-quality food or carbone-deficient water, no high-quality bacteria can evolve. The life-forms developing from these less highly-organised raw materials consume the body, originally brought to life and uplifted by high-grade bacteria. The correct composition of the blood and its inherent energies, which are determined by these metabolic processes, are of crucial importance. The decision to breed predators or beneficial organisms in our bodies therefore lies completely in our own hands - or in the hands and brains of specialists in agriculture, forestry and water resources. A certain uniformity prevails throughout Nature. Hence these symptoms appear in a similar fashion everywhere, as is evinced by other forms of vegetation. The errors that have been made must therefore take effect universally and must therefore provoke a general decline.

The inner material content of water is also crucial to the height of the groundwater table. As revealed by springwater rising vertically to the top of mountains, the inner energies in mature water become so powerful that they are able to overcome the inherent weight of water-masses, if aquifers are properly formed and not too large in cross-sectional area. Experimental proof of this is easy to produce, as the photographs in Figs. 7a, 7b & 7c show. The sinking of the groundwater table is above all a result of metabolic disturbances in the groundwater. Consistent with this phenomenon is disturbance in the circulation of blood in our bodies and the movement of sap in plants.

An Experiment

In *Our Senseless Toil*, Viktor briefly describes a 24-hour experiment using very simple everyday laboratory equipment (see *Living Energies*, p.132). This was designed to show the dynamics of true springs, the diurnal fluctuation in the height of the groundwater table and of sap in trees – Ed.

A small and unassuming experiment reveals a great law. Take a vessel, fill with sand, insulate the sides and the base from the effects of external temperature. By placing ice at the bottom of the vessel a temperature of +4°C (39.2°F) can be produced artificially - conditions will be created which are to be found inside the Earth.

Into the vessel thus prepared, insert a U-shaped glass tube into which is placed a quantity of pure quartz sand, which is almost chemically neutral. This sand should then be infused with saltwater. Both legs of the U-tube should then be filled with good water, enriched with absorbed and dissolved carbones matter which has not been exposed to sunlight.

On the open ends of the U-tube place two glass caps, onto one of which are braised two capillary tubes, and onto the other four capillary tubes. This must be very carefully done to ensure that the openings of the capillary tubes are not blocked in the process of being fused to the U-tube.

When this has been done, expose the surface of the sand bed to the Sun's rays. When the water reaches the anomaly point of +4°C (due to the ice-cooled sand bed at the bottom of the U-tube), and when the surface of the vessel reaches a temperature of about +20°C (68°F) as a result of the Sun's heat, then the water, known to attain its *greatest density and weight* at +4°C (39.2°F), begins to lose its equilibrium and rises up one leg of the U-tube if both legs of the U-tube are connected to the capillaries with suitably tapered unions.

If air is now entrained through the inlets on either side of the U-tube, as occurs for example with boreholes or wells driven into the Earth, the higher-rising water-column on one side sinks, and the water levels itself out in both U-tube legs, in accordance with the Law of Communication. If both side-

An Experiment

Fig. 7a (left): The constant pulsation in the capillary tube. There is no condition of equilibrium in nature.

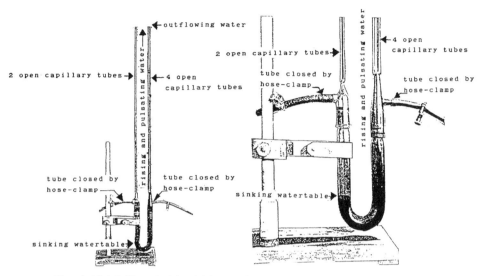

Fig. 7b (right): The principle of rising sap in the tree and circulating blood in the body.

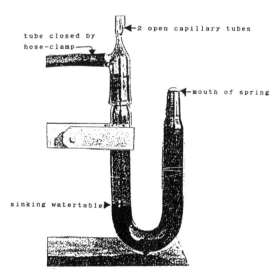

Fig. 7c: The principle of the mountain spring.

inlets are once more closed to atmospheric influence and the cold of the surroundings again begins to take effect, then after a while the water will once again begin to rise.

Why does the water subside as soon as it comes in contact with the entrained atmosphere?

If both openings are once more closed to atmospheric influence and the cold of the surroundings again begins to take effect, then after a while the water will once again begin to rise.

At night the process is reversed. In the capillaries in which the water previously rose a state of rest exists due to the effect of light and heat, whereas in the other capillaries the water now rises. The rising product of the alternating processes of equalisation exactly represents the different phenomena of night and day.

This simple experiment shows us why the substances that rise in plants during the day are different from those which rise during the night, and why the various types of blood circulate in our veins. Moreover it reveals many things to us about the secret of Life and its coming into being, which can only be achieved through contrasting conditions of heat and cold.

At the same time, this experiment also demonstrates the stupidity of the purely mechanical and thoroughly one-sided activities we call *work*, and how little we are aware of the underlying conformities with the laws of Nature and their processes. It would be beyond the scope of the matter at hand to explain all the necessary details and preconditions for the success of this experiment.

The development of all Life and the associated formation of structures is not merely a process to do with heat, as is assumed today, but also to do with cold - for Life can only be born out of differences. It is not possible to describe the subtle differences in the processes of decomposition and transformation that take place during growth, or which are necessary to transform such energy-bodies as coal, metals, minerals and elements and their compounds. It would likewise take up too much space to elaborate all the ways in which it is possible to accumulate dissociated energy-particles and coalesce them into an immaterial body.

One thing can be stated however: our learned scientists should quietly give up all ideas of violently splitting the atom in order to obtain free energy from the matter thus released. These attempts are both purposeless and absurd. Nature shows us in every blade of grass how it can be more simply and intelligently achieved.

Water Supply and the Mechanical Production of Drinking Water

[From *Our Senseless Toil*]

As well as campaigning fiercely for water to be allowed to flow naturally instead of being artificially constrained, Schauberger was adamant about the vital importance of using the correct materials for the pipes in which it was conducted. From his own research, he determined that the alarming increase in the incidence of cancer in new suburbs in and around Vienna went hand in hand with the expansion of the water reticulation system in which unnatural materials had been used. - [Editor]

Water supply

If we study the water supply system of the ancient Romans, we can observe from archaeological remains that when their towns were originally founded great trouble was taken to deliver the necessary water to the place of use in wooden pipes and in conduits of natural stone. It was only later, due to the constant increase in the demand for water as the towns grew, that they hit upon the unfortunate idea of conducting bathing and drinking water in metal channels.

Where the use of wood was discontinued, the choice of suitable material for the pipes was determined by observation of the behaviour of coins of different metals, which were thrown into the springs for ritualistic purposes. The type that best resisted the various influences over the years was selected. According to the nature of the water many metals became thoroughly encrusted, whereas others were almost entirely dissolved. When water is supplied in long, iron water pipes serious material transformations can occur under certain circumstances which cannot be detected with present-day instruments, but which are of crucial importance for the character or the psyche of the water.

It is known that electrolytic processes - energetic processes - are instrumental in the formation of rust. These take place on the internal surface of pipe-walls through the action of carbon dioxide, which evolves as a result of changes in temperature and the presence of oxygen. The carbon dioxide released through the reciprocal effects of heat dissolves the iron in the pipes, in the process of forming ferric bicarbonates. If further quantities of oxygen are added as a result of excessive aeration of the water, then with the simultaneous onset of electrolytic processes the ferric bicarbonate will be converted into hydrated iron oxide (rust). This is precipitated from the water as iron ochre, causing a narrowing of the pipe diameter. This is because the volume of wet iron rust is ten times greater than that of the parent material.

As a direct consequence of these processes a certain amount of carbonic acid is removed. This was formerly contained in the water as an essential ingredient in the constitution of its psyche. Hence the psyche of the water deteriorates. The transformation processes that take place at certain temperatures and which lead to the formation of iron ochre as an end-product already have artificially-pretreated iron as a base material. Whatever natural character the ore lying deep inside the Earth may have possessed is removed as a result of smelting and the admixture of various ingredients. If in the process of forming hydrated iron oxide, solid components accumulate on the inner wall surfaces, then transformation processes take place in conjunction with a negative temperature gradient. These eventually result in retrogressive transformations and lead to the formation of a new inferior psyche which, in certain measure, appears to be associated with the iron ochre. The water, therefore, has not only lost its high-grade psyche by being conducted in iron pipes, but in addition has become endowed with a pernicious and second-rate psyche.

An especial danger arises through the frequent application of tar to the inner walls of iron water-supply pipes. This is done in order to inhibit the formation of rust. It is a well-known fact in medical science that the extremely volatile products of coal-tar distillation give rise to cancerous diseases in the body, which is why some water-supply authorities have prohibited the use of tarred pipes.

As often happens, water conducted in this manner is also impelled through turbines and physically smashed to pieces by the high rotational velocity of the blades. When it is discharged from the turbines and subsequently mixed with other water, serious damage is inevitably inflicted on the organisms and the surrounding soil to which such water is supplied. This treatment of the Earth's blood one can roughly equate with blood transfusions where any kind of blood is drawn off, stirred up with a whisk, blended indiscriminately with foreign blood and then injected into the body. A person treated in such fashion can become seriously ill and ultimately insane. The same thing must also happen when water treated in the above fashion is drunk over an extended period of time. The blood will be systematically destroyed. The physical and moral degeneration of those forced to drink such water constantly should indeed provide ample proof of the accuracy of what has been stated above. Even the spread of venereal diseases is above all to be attributed to far-advanced, enfeebled conditions of the blood.

If the abusive debasement of the psyche of water is to be avoided, then it is essential that the material selected for the supply pipe is not only a poor conductor of heat, but is also of a properly-formed organic nature. The capillary is the best model of an ideal water-conduit for the proper conduction and treatment of water in terms of its material composition, internal configuration and associated functions. The most suitable material is good, healthy wood. Artificial stone (such as concrete) on the other hand is almost as unsuitable as metal for the manufacture of conduits, because only materials of natural origin should be used for the conduction of the Earth's blood. To those who protest that wood is unsuitable for the reticulation system of a city because of its limited durability, it should be pointed out that good, properly-treated wood can actually last far longer than iron.

Circumstances permitting, and apart from any other special treatment, these pipes should be laid and surrounded by sandy, humus-free bedding material in order to avoid external destructive influences to which pipes laid in the ground are frequently subjected. The poor thermal conductivity of wooden pipe-walls inhibits influences detrimental to the water's inner metabolic processes. This considerably weakens dissociations that take place under a negative temperature gradient and at the same time retains the quality of the flowing water.[26]

The hydraulic efficiency of pipes constructed with wooden staves is actually somewhat greater than that of iron or concrete pipes. The frequently-cited fact that wooden pipelines are cheaper to install should also not be underestimated. In any event, as must be emphasised here, the types of timber currently cultivated by modern forestry are well nigh useless for this purpose: almost without exception today's artificial plantation forests furnish timbers that possess neither the properties nor the durability of timber grown under natural conditions. It is rare today to find forests in which humanity, as forester, has not interfered destructively. Yet there are still suf-

---

[26]Wells (springs or shafts) should be protected (sealed) against all atmospheric influence. No iron pipes, concrete or brick that have been burnt should be used, for all that fire has touched becomes polarised and degraded and absorbs waste matter. For water pipes it is best to use fir, pine or larch which has been felled under a waxing moon during the period when the sap-flow has ceased. The bark should be left on the timber. Where there is a risk of rapid decay due to the acid content of the soil, the pipes should first be wrapped in old newspaper and then bedded in sand. Good quality (clean) loam may also be used and the whole should be back-filled with normal earth. Such wooden pipes last just as long as iron. In any event, it is far better that the pipe should fall to pieces rather than that the body should imbibe de-energised water. Here there is no place for economy. If it is to be maintained in a healthy condition, water must flow in an environment which is naturally-grown, otherwise its energies are discharged. Water is an organic magnet, and at the same time a transformer, a receiver and a transmitter. It is the mediator or accumulator of growth, which mediates life, but which (like a magnet) is discharged if it becomes warmer en route to its point of use. This is why we are so invigorated and refreshed after drinking good water. - *Tau* No. 153 - VS

ficient remote stands of valuable timber untouched by contemporary forestry, to which the greatest attention must be paid if humanity is once more to be supplied with good, healthy water. Once a suitable timber has been selected pipes can then be manufactured which largely correspond to the necessary requirements.

Water can only conserve its pipe system, however, if its inner conformities with natural law are taken into account. These inner conformities prevail if the substances the water secretes, which serve to maintain and build it up, are able to fulfil their respective purposes. It need hardly be emphasised that the quality of the remaining sources of food will inevitably decline with the general deterioration of the water.

The capillaries in animal or vegetable bodies serve for the transport of blood or sap, and for the simultaneous and continuous build-up and maintenance of the capillaries themselves. Hence drinking water supply pipes must be constructed accordingly, otherwise unwelcome processes will occur which lead to the destruction of the capillaries in the pipe-walls and to unwholesome metabolic processes in the water itself. These subsequently have the most detrimental effect imaginable on the human organism and on other bodies.

We find something akin to this in all waterways. Experience teaches that rivers seldom attack their banks if their inner conformity with natural law has not been disturbed. On the other hand there are no artificial bank-rectification measures that have proved effective long-term, which can withstand the destructive force of water whose natural flow has been impeded. The reasons for this lie in erroneous methods in use today which do not influence the water itself (which is what really matters), but which try to control it by its *banks*. Similarly it is of supreme importance that the composition of the walls of drinking water pipes is suited to the natural, inner functions of the conducted substance, otherwise in the first instance the supply pipes will be destroyed and in the second, it would lead to the destruction of the system of blood-vessels in the body. As a result dangerous metabolic disturbances would occur, which are largely responsible for the increase in cancerous diseases.

If the place where a spring is tapped is far removed from the point of use, it is only possible to maintain the character of the water (and then only partially) by taking very specific precautions. In no way can this be achieved with present systems of water reticulation, which are dictated purely by shallow and superficial reasons of profitability and expediency. The only case where slightly more care has been taken in the selection of the pipe material is in the transport of mineral water, wherein the emanations almost leap to the eye. Furthermore in order to satisfy the demand for water, springwater is often supplemented with immature groundwater which still lacks the requisite content of high-grade carbones.

As things now stand, if water warms up on its long journey through pipes, which unfortunately are today made mainly of good thermal conductors, then both carbones and oxygen in the water become more aggressive. The untoward effect of this development is revealed, inter alia, in the characteristic corrosion of turbine blades. The oxygen content enables embryonic bacteria, represented by organic matter in the water, to develop into bacteria proper. Processes identical to those taking place in the water itself must also occur, if such carbon-deficient and oxygen-rich water succeeds in entering the body. Under suitable temperatures it will likewise inaugurate transformation processes in the body's prescribed substances. These are not body-building processes, but manifestations of decay. Under these circumstances the consumption of such water will become one of the major causes of the scourge of the twentieth century - cancer.

The Consequences of Producing Drinking Water by Purely Mechanical Means

The production of drinking water by mechanical methods alone also leads to unpleasant surprises in many areas lying close to the sea. The conditions for equilibrium between layers of fresh groundwater and underground seawater were exhaustively studied by Badon Ghijbens and later by Herzberg. In this case we are concerned with the problem of the hydrostatic equilibrium between two mixable fluids of different specific weight.

In his 1911 paper entitled "Contribution to the Hydrology of Northern Holland"[27], Wintgens writes about this as follows:

The specific weight of fluids 1 and 2 are G_1 and G_2 respectively, and the difference in height between the surfaces of the two fluids after the establishment of a state of equilibrium is H metres; in this case the interface between both fluids will lie at a depth h_1 where:

$$h1 = \frac{G_2}{G_1 - G_2} \cdot H \, m$$

below the overall surface of the fluids. Deriving from this equation, the calculated maximum depth of ground-water $h_1 = 42 \times H$ m, on the assumption that the specific weight of freshwater $G_2 = 1$ and saltwater $G_1 = 1.024$.

In Norderny's example, as quoted by Keilhack, the surface of the fresh-water lies between 1 and 1.5 metres above sea-level. By calculation this value of H of 1.5 metres would correspond to a ground-water depth of $42 \times 1.5 = 63$ metres. The actual depth of the ground-water was determined as lying between 50 and 60 metres.

[27]Beitrag zur Hydrologie von Nordholland'. - VS

If the freshwater-table is now lowered through the excessive extraction of water with large pumps, which also reduces the value of H, then the boundary-layer between fresh and salt water will be displaced upwards until it finally reaches the level of the suction-head of the pump, with the result that the salinity or the chlorine content of the drinking water increases until it becomes undrinkable.

These physico-mechanical processes are also augmented by the metabolic processes taking place between freshwater and seawater. Every new borehole driven into the Earth facilitates the penetration of oxygen into the boundary-layer between these two types of water. The condition of the temperature gradient between the surface of the freshwater and the underlying boundary layer will also be altered. The combined effect of these two components is such that the water's inner buoyant energies, which would normally maintain it at a certain level, are likewise reduced.

In this connection attention should be drawn to the salination of many mountain lakes, which is ultimately attributable to the activities of hydraulic and hydro-electric engineers. First of all, through the wanton clearing of forest the rivers were robbed of protection from Sun and heat afforded by the leafy canopy of the trees. In addition water-courses were subsequently subjected to regulation by mechanical means alone. Both events produced higher concentrations of oxygen in the water, which then sought out the coarse and fine carbones in the channel body, dislodging them from both bed and banks. Once this water reaches deeper and cooler lakes where the now-aggressive oxygen is concentrated and if the water is no longer able to retain the quantities of the now-dispersing carbones in suspension, the precipitation of salts then follows and freshwater is transformed into seawater.[28] The reverse process happens at great depths in the sea, where strong concentrations of high-grade, complex carbones can eventuate. There the water is not only fresh, but also develops a highly potent negative charge, which under certain circumstances can trigger off violent electrical disturbances in the depths of the ocean.

Deep-Sea Water

Were our learned scientists to investigate deep-sea water more thoroughly, they would discover that the material composition of the air absorbed by it differs substantially both in quality and quantity from that contained in extreme surface water. This fact also explains why deep-sea fish can glow and impart electric shocks. In its fundamental structure, the air absorbed by

---

[28]This perhaps explains the formation of marinectic lakes in which water is salty in the lower part and fresh in the upper part - Ed.

deep-sea water exhibits a similar composition to that still found in a few isolated high altitude springs. It is the high content of physically-dissolved carbonous matter and the lack of oxygen, coupled with the simultaneous exclusion of light, that gives this water its peculiar character.

Seawater at great depths cannot absorb gases by diffusion or convection. Therefore where the oxygen has also been consumed by living organisms, such seawater can actually be locally completely devoid of oxygen - or even fresh. Because the carbon dioxide content of the atmosphere is less over the sea than over the land, the conclusion can be drawn that the surface of the sea also absorbs carbon dioxide directly from the atmosphere.[29]

Deep-sea creatures can be distinguished from their counterparts in shallow waters by their size, their strangely-constructed eyes, the different consistency of their bodies and, to a large extent, by their particularly original body-shape. The external environment stamps each individual with its own characteristics. Hence there are certain contradictions which can only be explained by understanding the nature of the water in which these organisms live.

One would tend to think that, owing to the water-masses pressing upon it, an organism living in the deep sea would have a correspondingly more strongly-built body. However, in contrast to the fish with robust skeletons and strong muscles found along the shoreline, deep-sea fish have extremely delicate, paper-thin, almost weightless skeletal frames. The fact that these creatures burst open when they are brought up from the depths is also attributed to the physical structure of their bodies. This purely mechanical explanation is a serious error. Just as the organisms brought up from the deep sea regularly explode, the same occurs with water raised from such depths. It will warm up relatively quickly with the addition of the requisite quantity of oxygen and highly complex carbons (such as oil) or will burst its container if this is sealed.[30]

A great many natural phenomena occurring in the depths of the ocean could easily be explained were the experts aware of the inner nature and character of deep-sea water. This holds particularly true for the phenomenon of ebb and flood, whose true nature will be described in a later chapter.

For the same reason, our energy-technologists would abandon contempo-

---

[29]See Dr M P Rudzki: *Physics of the Earth.* - VS

[30]*"Above all it is the high content of physically dissolved C^e-substances and the deficiency of O (oxygen) under the exclusion of light that endows this water with its distinctive character. With the addition to seawater of appropriate quantities of O and primitive C^e-compounds, such as oil, it quickly warms up and explodes its container. Electrical energy can be obtained directly from deep-sea water with the aid of simple apparatuses."* Paragraph 7.1.2 from Notes and Letters by Viktor Schauberger published in a special edition of *Mensch und Technik* (Vol. 2 - 1993) - transcripts and descriptions from the 1941 notebook of the Swiss, Arnold Hohl about Viktor Schauberger's activities from 1936-1937.

rary methods of generating electricity if they but knew that this can be obtained directly from the deep sea by means of the simplest apparatus. Contemporary apparatuses and instruments would rapidly become obsolete because humanity has no need to go to such lengths to obtain light, heat and other forms of energy - it could be obtained in any desired quantity almost without effort or expense.

The Conduction of the Earth's Blood
[From *Our Senseless Toil*]

Before we move on to a description of the correct construction of a water-supply pipe, a further example should be given which should make the principle of correct water conduction obvious.

If the blood-vessels of a snail are examined, two differently-coloured systems of blood-vessels are evident. The blood flowing in the outer system of vessels is lighter and in the inner system, darker. The composition of the blood in the outer system is distinguished by a greater oxygen content and is substantially different from that of the inner system, which exhibits a higher content of carbones. Investigations further demonstrate that suspended matter is concentrated in the middle of the capillary cross-section, whereas dissolved matter congregates more towards the periphery. In addition, if the blood-flow is deemed to be moving along a straight line then the velocity of forward motion is less at the periphery than at the centre. In this regard, however, note that this difference in speed is only illusory. The forward motion of the inner particles of fluid only appears to be faster than the outer blood particles. This is because the latter must describe a path roughly corresponding to a double-helical motion - a spiral motion within a spiral - whereas in the main the inner blood particles appear to perform a simple spiral motion.

The second component of the double-spiral movement described by the blood corpuscles of the inner system cannot be observed, because the line of the second spiral is an energy path imperceptible to the eye. This has a much higher meaning, for we are here concerned with processes of qualitative psychic enhancement, the raising of the psyche to a higher energetic and immaterial level. This not only influences the character of the blood, but in the course of further development also affects the character or psyche of the respective organism.

In contrast to customary methods of investigation, the conviction is slowly gaining ground in various fields of research that the object under examination should be decomposed into its constituent parts to enable the study of its very smallest aspects.

The frequently-mentioned `material transformations in water` are outwardly identifiable through the pulsation of water. Hydraulics is only aware that this decreases with increasing velocity and is intensified with increased roughness of channel wall-surfaces. Water is therefore invested with a certain inner vitality and a decisive role as it rises up in the capillaries in concert with the supply of necessary formative substances.

On many occasions I have stated that the rising of sap in trees cannot be explained by physical factors alone - such as the effect of external air pressure. Its explanation is to be found in ongoing metabolic processes in constant pulsation in every cell of the tree. It is therefore a result of the vital activity of the capillary tree-cell. Professor Kurt Bergel of Berlin came to similar conclusions in relation to the activity of the heart and the blood in animal life. He rejects the current view that the motor - the heart - is supposed to pump blood into all parts of the body. On the contrary this work is performed by the millions of highly active capillaries permeating the body. This fluid-raising capillary force is only effective up to a certain height. An external aid is therefore necessary. This Bergel demonstrated with a small experiment. He stood the base of a bundle of hair-thin tubes in water and lightly and regularly tapped it at the top, causing water to flow continuously out of the upper ends of the capillaries.[31]

In his view health and disease are primarily dependent on the faultless or disturbed activity of the capillaries. Professor Bergel furnished definite proof of this in his investigations of a bird's egg. After being incubated for only a short period of time a small red spot appeared on an egg, which on closer inspection proved to be a drop of blood. If the egg is incubated further, then a network of arteries can already be distinguished on the skin of the yolk-sack. Rhythmical pulsations can still be detected just before it cools off.

The Double-Spiral-Flow Pipe

Both in cross-section and longitudinal section the double-spiral-flow pipe satisfies all the criteria necessary for a water-supply pipe, if it is to convey healthy water to the place of use. By means of a system of vanes made of precious metal arranged on the inner surface of the pipe walls (see figs. 8, 9, 10 & Patents Nos. 134543 & 138296 in Appendix), the water-masses are conducted along a double-spiral-flow pipe in such a way that the movement of

---

[31]From *Implosion* Magazine No 5, p.17: "Contemporary medicine ascribes the major role in the circulation of the blood to the activity of the heart, this barely fist-sized, muscular sack with an output of 0.003 hp. The human heart 'pumps', so they say, 1/10th of a litre of blood into the arterial network about 75 times a minute; 10,000 litres per day, almost 4,000,000 litres per year. The work of the heart, according to modern medicine, would be sufficient to raise a weight of 40 tonnes, 1 metre high."

Walter Schauberger.

the individual filaments of water at the periphery takes the form of a secondary helical motion along a primary helical path (see fig. 11).[32] Through this arrangement both centrifugal and centripetal forces evolve simultaneously in the cross-section of the pipe, which convey bodies heavier than water down the centre. Bodies lighter than water are impelled towards the periphery.

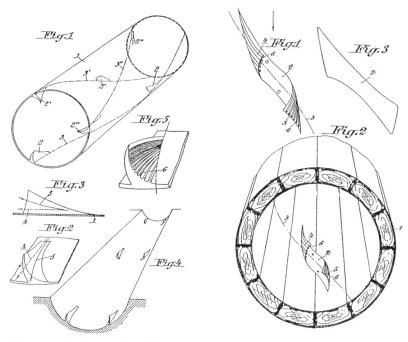

Fig. 8: Patented double-spiral flow pipe. (See Patent No. 134543 in Appendix)

Fig. 9: Patented wooden double-spiral flow pipe. (See Patent No. 138296 in Appendix)

Water-masses conducted in this fashion are slightly warmed through the interplay of mechanical forces of friction on the vane-surfaces, leading to the separation of oxygen in the inner region of the pipe and its subsequent concentration at the periphery.

At the same time as the oxygen is ejected, *all the bacteria* migrate towards the periphery as well, since their living conditions in the more central part of the cross-section have now become unsuitable. In company with the bacteria, all the water-polluting particles are also dispatched towards the periphery of the pipe. Thus the water is easily and simultaneously purged of suspended matter.

[32] See also the translations of Austrian Patents Nos. 134543 and 138296 in the Appendix.- Ed.

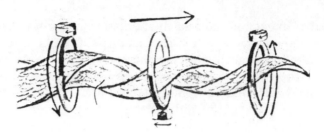

Fig. 10: Viktor Schauberger's portrayal of the double-spiral longitudinal vortex

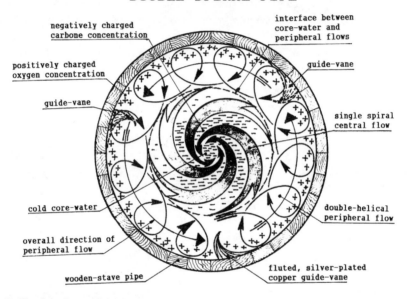

Fig. 11: The flow dynamics of the double-spiral pipe

Once bacteria have transferred to the peripheral zone in search of the required oxygen, and after a certain period of time in water completely cut off from outside influences, they are overwhelmed by a localised concentration of oxygen. In this way precisely those pathogenic bacteria susceptible to an excess of oxygen are advantageously eliminated, whereas non-pathogenic bacteria which are not harmful to human health, but in many cases are actually beneficial, are to a certain extent retained. At the same time as the content of absorbed oxygen is separated from the carbones contained in all water, the inner core of the water surges ahead in a simple spiral movement (vortical movement along the longitudinal axis), because the surface tension of the water becomes physically reduced as a result of the above-mentioned separation of oxygen from the particles of carbone.

The physical reduction in surface tension results in a mechanical acceleration, leading to the self-purification and energetic *charging* of the centrally-accelerating water-masses. On the other hand this charging of energy gives rise to further processes related to the overall equilibrium between the heavy, centrally-accelerating bodies and the energy-rich water. With the ensuing simultaneous cooling solid particles are separated and are again directed towards the periphery. There they combine with oxygen and are reunited with the centrally-accelerating water in the form of additional energies. Those particles of matter not drawn into the centre will be pressed onto the surface of the pipe walls by the prevailing mechanical pressure, there to combine with the raw materials from which the timber was originally formed. Thus they seal the pores of the wood, which in this way *becomes more durable than iron.* Once again we are here concerned with a natural process whose active principle is operative in the formation of all capillaries. The capillaries not only construct themselves, but also protect themselves against harmful influences.

As a result of acceleration of the entire body of water peculiar to the double-spiral-flow pipe, greater quantities of water can be conveyed than in an ordinary smooth-walled pipe and, due to the efficacy of the oxygen, extensive self-purification and self-sterilisation of the water occurs which *constantly increases in quality* through the uninterrupted build-up of energy as it moves along its path. The reason for this is as follows: as they accelerate, centrally-conducted water-masses are simultaneously cooled, with the result that gases evolving from the carbones become concentrated in the flow-axis, where the lowest temperatures reside. This concentration decreases towards the periphery. The oxygen on the other hand is concentrated around the periphery of the pipe, reaching its most aggressive state at the interface with the warmer pipe-wall, giving rise to mutual interactions between the two basic substances from the periphery inwards. This subsequently leads to the aforementioned interactions which qualitatively enhance both water and wood.

In the course of time the relative spacial distribution of the more central flow of water and the interactions at the surface of the pipe walls arrive at a certain state of equilibrium. These processes then cease - the water is now *mature* and both wood and water have become almost immune to harmful outside influences. Whereas oxygen is located in the *peripheral zones of the pipe*, the *free* particles of carbonic acid congregate in the *boundary zone of the inner core of water* as a result of the water temperatures prevailing there. The carbones contained in the water, in bound form, necessarily accumulate in the central axis, which is predominantly saturated with carbones. By arranging the in-built, specially-shaped vanes in a particular way, aggressive particles of oxygen on the boundary layer of the outer edge of the inner core of water, are brought into continuous and direct contact with the most aggressive carbon dioxide, resulting in a continuous generation of energies. These are drawn further towards the centrally-accelerating water-masses, due to the decrease in temperature towards the central axis of the pipe.

Accordingly two types of circulation are created in the cross-section of the pipe: the mechanical circulation of the water and the *counter-circulation* of those energies that evolve when aggressive particles of oxygen encounter free carbon dioxide. This circulation of energy manifests itself in the form of a continuous electro-dynamic process. In this instance it does not take place at the walls of the pipe, but at the boundary zone of the water's inner core, resulting in the *qualitative uplifting of its physical, material, energetic and immaterial attributes* - but not in the destruction of the pipe walls.

These double-spiral-flow pipes also convey matter heavier than water down the middle of the pipe and at the same time ennoble and refine it, so that oils of inferior quality, for example, will be improved during flow. After smelting, iron ores transported in this fashion yield a higher-grade iron, because in the process of being transported, the oxygen in the ore is consumed in the formation of new carbone compounds (reduction processes), which then contribute towards the materially higher composition of the carbone - iron.

The Pulsation of Water

Life takes place in three spheres:

1. in the Carbonesphere;
2. in the Atmosphere;
3. in the Stratosphere.

These various spheres are inter-connected through the agency of water. On the other hand water's various states of aggregation form the bridges for the formation and transformation of the basic elements it transports, which are thereby transferred from the stratosphere into the interior of the Earth and *vice versa*.

Apart from the purely mechanical cycle of physical water, there is another form of circulation which operates in the opposite direction - the `energy cycle`. Here two contrasting movements are involved: the upward movement of carbones with the carrier water and the downward movement of oxygen. At the point where the paths of these mutually opposed currents intersect, energy is freed. Owing to the constant variation in the length of night and day, these energetic interactions cannot arrive at a state of equilibrium. So shifts in individual micro-climatic conditions must constantly occur, resulting in continual modification of the qualities and quantities of basic elements.

The outcome of this continuous reciprocal interaction is the metamorphosis of the different forms of water in each individual zone and the constant transformation of the various species of vegetation in which water, ceaselessly moved by this interplay of forces, wends its onward way. This inner equalisation of forces is to a certain extent inhibited by the effect of the self-weight of the water. Through fluctuations in the magnitude of the component forces, water particles are constantly made to rise and fall. In other words the *pulsation* of the water inevitably occurs.

Every new entity and all growth evolves from the smallest beginnings. Further development in the early stages can only be accomplished if the cycle in the interior of the Earth proceeds correctly. In the natural order of things, every higher form of vegetation is built up from the lower forms preceding it. The carrier of the necessary elements and the moderator of life-processes in the vegetation root-zone is groundwater, whose movement is triggered by a drop in temperature, which on its part is caused by inner metabolic processes taking place in the here-decisive groups of basic elements. The impulse for the movement of water is hence a product of interactions between the opposing elements it contains, the necessary resistance for which is supplied by the water itself. This resistance to the interaction between the carbones and the oxygen further results in continual fluctuations in temperature, which impart an additional impulse to the movement - the pulsation of the water.

As it moves along its path water dissolves salts at one moment, transports them at another, deposits them at a third and creates and transforms energies. The essence and purpose of these eternal metamorphic processes is to build up and maintain various forms of vegetation and physical bodies, which on their part represent bridges by which energies are created and maintained. The potential differences that are always present between the inner temperature and the outer ambient temperature are none other than different patterns of force which both terminate and reactivate the circulation of water.

The two forms of evolution are therefore of material (physical) nature and of immaterial (spiritual) nature. All that exists, be it a stone, a plant, an animal, a human being, a planet or the Sun, represents an organism possessing both body and soul. Every ray of light and heat requires a physical form in which it can manifest itself and evolve. Every body requires an inner energy which creates and transforms it. When a body decays, then the energies that initially created it are again freed. They are never lost - should they lose their habitat with the decay of the body, then they willingly coalesce with the water that eternally circulates in, on and over the Earth and transmute it into to a new life-form. Therefore wherever we look there is life, eternal creation and transformation. Should we look into apparent emptiness, then a sea of spiritual life, of past and future generations stares us in the face.

Every material form of vegetation always has its immaterial counterpart in light, heat and radiation. Every change in sphere changes the overall internal and external conditions, alters the weight and inner intensity of the radiation of the physical substance - water - and thus the direction in which the carrier of life moves. Disturbances to the inner and outer conformities with natural law lead to the disturbance of the very foundation of all life's creative processes. The disappearance of water or its material transformation is a very serious warning sign, for if its inner structure and composition are changed then its character and hence the character of all forms of life, including humanity, will also be changed.

Regression in the quality of the various forms of vegetation, qualitative deterioration of the highest plant organism - forest - and the increasing physical and moral decadence of humanity are simply the logically-consistent symptoms of disturbance of the physical constitution of water and disruption of the geosphere brought about by humanity's subversive activity in the organism `Earth`.

The religion of the Chinese forbade any kind of intrusive interference with the Earth. Even the construction of railways in China encountered stiff opposition in religious circles. In the light of all that has been stated so far it can be seen that the culture of the Chinese, which has endured longer than that of any other people, is no accident. It owes its existence to the fact that the carbone-sphere remained untouched for a long period of time. China's decline was inevitable the moment the Chinese began to adopt the manners, customs and technological achievements of the western world. Another horrendous example concerns developments in Russia, which took only 15 years to catch up with what other nations had taken 100 years to achieve - famine.

What we are experiencing today is no crisis. It is the dying of the Whole - the qualitative, physical deterioration of all organisms initiated by the disruption of Nature's water-balance. In step with this proceeds the moral, intellectual and spiritual collapse of humanity. This collapse is already so far advanced that despite all the warning signs, people are still unaware of the gravity of the situation. Behaving far more cruelly than animals, people see their ultimate salvation in decimating the mass of humanity with weapons which - together with the banners under which our children are supposed to bleed to death - our priests actually bless.

The decision - whether we take the latter road of collapse or save ourselves at the eleventh hour from our own self-mutilation - lies only with ourselves or with politicians and men of science who take upon themselves a *truly appalling responsibility*. They do not consider the seriousness of the situation and are unable to provide any really effective assistance in salvation - and out of selfish interest they continue to adhere firmly to their present point of view.

Healing Water for Human, Beast and Soil

As long as humanity refrained from interfering with Nature's interdependent organic functions, and as long as Mother Earth could still supply her blood - water - to the vegetable kingdom in a healthy condition, there was no necessity to contemplate how wholesome water could be produced artificially in the same way that it is naturally constituted inside the Earth.

Today, where almost all healthy springs have either dried up, or the water has already been intercepted at its source and delivered to urban areas in wrongly-constructed water-mains, the soil and the entire animal kingdom have to rely on worn out, stale and consequently diseased water. Even quite immature water (full of inferior, less complex substances, and torn from the womb of the Earth), or health-endangering surface water (sterilised with chemical additives) must be supplied for human use. It is therefore high time that we discover ways and means of protecting human, beast and soil from the decay that must legitimately follow if the Earth dies of thirst - as a result of the internal decomposition of water arising from current economic measures and industrial practices.

Nature alone can and should be our Great Teacher. If we wish to regain our spiritual and physical health, we should not simply rely on secondary mechanical or hydraulic phenomena. As a first priority we must see to it that sublime conformities with natural law are thoroughly investigated - conformities which govern the ways in which Mother Earth prepares her life-giving fluid and the means she uses to conduct it to the point-of-use. Once we have unveiled this secret, if we faithfully copy what has been tried and tested over millions of years, then we are on the right track. Only then can we intervene analogously in Nature's vital functions and harvest an over-abundance of the best and most noble fruits that Mother Earth has created and maintained in countless varieties with the aid of healthy blood. In order to penetrate the great mystery surrounding the origin of all life, we must take an interest not only in our living space, but in the `above and below`, in which water pursues its eternal cycle in obedience to a great and immutable law.

It may be impossible for us to observe the wonderful processes in crystal-clear water with our eyes, and likewise impossible for us to accompany water on its mysterious path above and below the Earth. Nevertheless an indirect, inductive way still lies open to us to research those things we cannot see but which we absolutely must know about, if we wish to remain healthy and hence to serve the purpose of life: *continuous creation*.

Up to now all that humanity has ever done has been to commit crimes against Mother Earth. In so doing not only do we inflict grievous injury on ourselves but also on the natural environment. With endless patience she has passively suffered humanity's interference and intrusions, motivated by greed, avarice and ignorance. However as a result of continuous ransacking and thorough ventilation of the Earth, inner decomposition of her blood is taking place and with it the dying of the soil that feeds us.[33] Not only have pumps been attached to the inner circulation of water, wrenching it prematurely from the Earth's womb, but also the water flowing over the Earth's surface has been ruined through senseless regulation of water-courses. As if this were not enough, we also cut down Mother Earth's forest or destroy it organically, and now our own head is finally on the chopping block. This had to happen in order to bring humanity to its senses and to the understanding that nothing in this world goes unpunished. Ultimately every foolish interference with Life's wondrous workings - Nature - must exact its vengeance on humanity itself.

The fable of a former paradise is no figment of the imagination. Although our ancestors may indeed have been engaged in a constant struggle for survival, their lives were still relatively carefree by comparison with the present era. But what will it look like after a further generation, if it continues to go downhill at its present rate? What future will our children have to face if no way can be found to stem this dreadful, festering tide? Today we are already confronted by events that must shake every thinking person to the very core of their being. What purpose is served by continuous self-deception or by deluding ourselves in the foolish hope that somehow things will improve by themselves? If we wish to make life enjoyable and beautiful

---

[33] "The intensive ransacking of water, its analysis and measurement with every possible method of investigation and observation in water research organisations, is without limit. The 'water-corpses' thus examined will never be able to reveal their laws to the light of day. It is only through the ways in which moving water expresses itself that one or two conclusions can be drawn and something inferred. The more profound conformities with natural law, however, lie hidden in so-called darkness deep within the organism of the Earth and in bundled form within the various forms of organic life. The blood of the Earth pulsates in deep and dark pathways according to its own ur-immanent law, and it is this law that governs all life. In contrast to the assertions and definitions of chemistry, many deliberations and observations of Nature have led inter alia to the surmise that water cannot be unequivocally defined by the formula H_2O. Rather, water is a component containing a specially structured combination of carbone and hydrogen. On exposure to the atmosphere or the influence of daylight, water becomes 'oxidised'. The visible manifestation of this oxidation is the release of carbon dioxide (CO_2), which is accelerated commensurately through the influences of strong light and heat."- VS - *Implosion* Magazine, No. 103, page 30.

again, then we must apply the lever where life begins. The origin of life - the *ur*-substance - is water, which is the guardian of the secret of all becoming and evolution. This secret will only be unveiled once we have come to understand the innermost nature of water.

In exactly the same way that a ripe apple falls to Earth from the tree, water rises out of the Earth of its own accord when it is mature. It matures when it has so transformed itself internally that it can and must take leave of Mother Earth by overcoming its own physical weight.

While correct methods of spring-capture cannot be addressed in great detail here, reference should nevertheless be made to the skills of the ancients. These skills were either lost or had to yield to worse practices. Where possible, the Romans tapped their springs in such a way that at a certain height above the mouth of the spring they placed a cover in the form of a thick stone slab, carefully levelled and smoothed, on the sloping face of natural rock. Having been completely sealed all around the perimeter with driven wedges, a hole was then cut in the stone slab into which the outlet pipe was inserted and secured, so that no entry of air was possible. In spite of and because of their simplicity, all methods of spring-capture in those days were more mindful of the nature of water than contemporary systems. Apart from other serious errors, contemporary systems have also frequently destroyed the conditions of water circulation and metabolism between the spring and its surroundings, principally by over-extensive building works and by disturbances in the vicinity of the spring caused by the use of lime, cement and metal pipe-fittings.

The following statements should not be taken as a recipe for the production of healthy water. It should only be stated here that even in this area a thinking person can make good the sins of his or her ancestors, and is capable of producing good, healthy water in the same way that the Earth does.

It is quite obvious to us that a mighty tree can ultimately develop from a healthy seed planted in the earth. It is thus equally understandable that only ripe and healthy water can produce healthy fruits. In the same manner that a seed in the moist earth requires heat and cold, light and shade, and the energies associated with them, exactly the same applies to water. Water has an equal need of these opposites in order to build itself up and reconstitute itself internally. The very reason water wends its long way through the universe is to maintain and gather these opposites. In every drop of water dwells a world of possibilities. Even the Divine has Its abode in every drop of water. If we destroy water, if we remove it from its cradle of the forest, then we stupidly rob ourselves of our most prized possession - our health. With it we lose our place of birth - our habitat - as well. As restless as water which has been wrested of its soul, we too must once more take to the road. Wherever we alight, decomposition, unrest, ruin, poverty and privation soon begins.

However, if our work is to become a blessing instead of a curse, then we must content ourselves with living off the *interest* and the superabundant, ripened products of the Earth's capital. We must never live off the substance of the Earth directly. Water supplies this interest in such a valuable form that we could do without all the rest and live off her surplus alone, taking only what is ripe, once we understand how the Earth manages her household. We still have time and we still have water. If at long last we finally take proper care of this giver of life, then all will right itself again automatically!!

Good high-grade springwater differs from atmospheric water (rainwater) in its inner material contents. Apart from dissolved salts, high-grade springwater possesses a relatively high quota of gases in *free* and *bound* form (such as carbon dioxide and carbonic acid). Up to 96% of the gases absorbed by good high springwater consist of compounds of carbone. Under the term carbone is here to be understood all the carbons of the chemist, all elements and their compounds, all metals and minerals - in a word, all substances except oxygen and hydrogen.

Atmospheric water (rainwater, aqua destillata, condensate) or surface water exposed to strong aeration and intense light influences exhibit a comparatively high oxygen content, almost no salt (or only less-complex forms), little or no free carbon dioxide and bound carbonic acid, and an absorbed atmospheric gas content predominantly of oxygen in *physically* dissolved form. The expression `physically dissolved form` here means a more highly-evolved solution (compound), comprising groups of substances not occurring in purely chemical forms of solution, and in which *energetic* processes are already actively involved.

Following from this, we therefore differentiate between water possessing a high percentage of *carbone energies* and water exhibiting a high percentage of *oxygen energies*. The former we will describe as *cathode-water$^-$* and the latter as *anode-water$^+$*. Cathode-water possesses a *negative* form of energy and anode-water a *positive* form. These energy-forms are characteristic of what we describe as the sphere, psyche or character of water. Accordingly, high-grade springwater bubbling out of the Earth possesses a preponderance of *carbone-spherics* - negative energy-forms or negative character - whereas rainwater coming from the atmosphere chiefly exhibits *oxygen-spherics* - positive energy-forms or positive character.

Apart from the necessary *isolation from light and air* and a capacity to absorb certain transformative substances (metabolic catalysts), atmospheric water infiltrating into the ground also requires certain lengths of path and periods of time in order to carry out the restructuring process correctly - to become inwardly *ripe*. Only mature and therefore healthy water can produce good fruits. In the same way that the seed requires heat, cold, light and shade and the energies associated with them for its development, so too does water in order to be able to build itself up and transform itself internally.[34] Water is ripe when its absorbed air contains at least 96% carbon-spherics, together with a quota of solid carbones associated with such a sphere. It is precisely upon this inner ripeness that water's excellence and up-rising or levitative force depends. The longer the path travelled, the more highly-organised and qualitatively higher-grade its inner energy becomes, provided that the appropriate transformative substances are present. The closer to the centre of the Earth, the more complex and aggressive the oxygen-groups infiltrating with the water become.

When atmospheric water infiltrates into the ground its oxygen content becomes concentrated as it approaches the geothermal low-point of +4°C (+39.2°F). All the carbones present above this boundary layer, which combine with the particles of oxygen as they approach it, are thereby restructured. Some of these rise upwards as nitrogen while others remain behind as salt crystals.

Such oxygen-charged water can therefore take with it none of the carbones previously brought up from the Earth's interior through the reverse process, below the boundary layer of +4°C. It must leave them behind in the vegetation zone. This vegetation layer is akin to a sub-depot which is continuously supplied with oxygen or carbones from above or below through these reformative processes. It is limited in depth by the geothermal neutral layer of +4°C. The water that sinks further beyond this boundary layer can only take with it those surplus or less-complex portions of oxygen which cannot interact or enter into a restructuring process (oxidation), for lack of the presence of suitably-organised carbones in the vegetation zone.

Due to rising temperatures with increasing depth in the interior of the Earth, which themselves are actually engendered by these interactive processes, the oxygen descending with the water becomes increasingly aggressive. This enables both the interaction and recombination of various grades of oxygen with carbones. These carbones have also become progressively less complex with increasing depth. Ultimately even coals (carbones in a solid state of aggregation) are decomposed and restructured when aggressive oxygen comes into direct contact with them under high pressure (which arises simultaneously because of this). Incidentally, we also find something similar in the transformation of foodstuffs in our bodies: this transformation takes place with the intake of water and air, and activates metabolic processes that condition life.

The higher these reconstituted and ennobled carbones rise towards the Earth's surface, the lower the surrounding temperatures become with the approach towards the boundary layer of +4°C. During this process oxygen

---

[34]Last two sentences from Spec. Ed. *Mensch und Technik*, Vol. 2, 1993. - Ed.

components of groundwater also become less aggressive. The higher-grade the carbones, the less complex the oxygen groups need be in order to complete the interaction, and *vice versa*. The relative position of the boundary layer of +4°C also varies, due to fluctuations in ground temperatures caused by the rising and setting of the Sun and the alternation of the seasons. Generally speaking, this layer lies higher by day and deeper by night. When assessing the causes of fluctuations in the groundwater table, the introduction of the already well-known concept of the saturation deficit is necessary, with which the relation between temperature and water-vapour content of the atmosphere is determined.

The climatic conditions of Central Europe are of moderate continental character and are distinguished by maximum rainfall in the summer months. However, this is associated with a corresponding increase in evaporation due to higher temperature - thus the saturation deficit will be greater. The annual rainfall distribution in Central Europe amounts to 9%-13% in the summer months and 4%-6% in the winter months. According to Mayer's findings (*Meteorologische Zeitung*, 1887), these values for rainfall distribution are to be compared with summer and winter saturation deficits of 3mm-7mm and 0.3mm-1.0mm respectively.

With equal levels of relative humidity, and with rises in temperature from -10°C to +30°C, the water content of the atmosphere can rise more than fifteen-fold. Only when data concerning the amounts of rainfall and the saturation deficit have been studied will it be possible to arrive at laws governing the fluctuations in the height of the groundwater table. Generally speaking, however, since at present these two meteorological components can neither be added together directly, nor cancel each other out, fluctuations in the groundwater table must therefore be dependent primarily on their reciprocal interaction. The possibilities for practically applying the here-decisive conformities with natural law encompass the effortless and almost costless raising of the deep-lying groundwater table in deserts.

Apart from the mechanical interplay of forces co-active in the raising and sinking of the groundwater table, another factor to be taken into account is the physical interaction - the absorption of portions of carbone elements and the binding of gaseous carbones, which are diffused (dispersed) through the water at a suitable temperature if isolated from light and external air. The highest dispersion of carbone groups is always present in the immediate proximity of a concentration of oxygen that occurs under these circumstances - which means that the water can complete its reconstitution and become internally mature.

The water lying above the boundary layer now further charges itself with carbones present in the vegetation zone of the Earth, using up more and more of its oxygen in the process. When a certain degree of saturation is reached following an increase in ground temperatures towards the ground surface in summer, it then has to release carbonic acid, which rises in the form of bubbles and mechanically assists in raising the water in the soil's capillaries. This interplay of forces is boosted by yet another physical energy-form - the oxygen-starvation of water over-saturated with carbones, which creates a negative pressure (vacuum), resulting in the raising of the water.

Good, high mountain springs do not gush out of the ground due to excess mechanical pressure (as has hitherto been assumed), but because of the effects of *negative pressure* (suction). In the final analysis these are due to processes of material transformation - combination of mechanical and physical effects related to the non-compressibility of water at +4°C. This explains the phenomenon of *high-altitude springs* that rise on mountain peaks or at great heights, which are caused to rise to the surface through the action of physical opposites.

When carbones, whose quality constantly improves the higher they rise, draw closer to the concentration of oxygen present in the upper regions of the atmosphere, the last remnants of the accompanying water crystallise under the low temperatures prevailing at this altitude. They descend with the oxygen again as microscopic particles of ice. Now moving without a carrier and continuing to rise, the remainder of the extremely diffused carbone particles ultimately reach the highest oxygen concentration of all - the Sun - and contribute to the organic, formative processes of the solar system. The reverse process takes place in the depths of the Earth, where carbone groups - coals - already compacted and concentrated, are decomposed under the influence of the most highly-aggressive oxygen.

The energies in the upper regions of the atmosphere, which evolve from the interaction between highly-complex carbone groups and less-complex quantities of oxygen, return to Earth again by way of radiation. Conversely, radiant energies that have been released in the depths of the Earth are drawn upwards. Gaseous hydrogen, which becomes denser as it approaches the Earth's surface, offers a resistance to the interactions through which these energies are transformed into light or thermal radiation. In this form they finally reach the Earth and contribute towards the organic build-up of various forms of vegetation. The nature of the processes taking place deep inside the Earth is such that their effects are projected in the opposite direction. Radiation, light and heat are therefore the counterparts of certain forms of energy evolving at the Earth's surface.

Vegetation (material bodies) is equally the result of the restructuring processes continuously taking place. Water is everywhere involved, and with its assistance the necessary interactions occur. Every change in form of vegetation hence inevitably leads to the modification of this inner transformation and development, to an alteration of climatic conditions - and thus to a change in the inner character of the world's blood, water. The properties or *character* of the Earth's blood is conditioned by the sum total of circumstances that have only just begun to be considered by our experts. The beneficial or detrimental influences of certain substances contained in water, such as chlorine, ammonia, manganese, iron, sulphuric acid and so on, will not be discussed here, since these are dealt with quite sufficiently in the relevant technical literature. From our point of view, we are primarily interested in the oxygen content and the carbon dioxide content in its various bound forms, including its salts.

In various publications it is gradually becoming evident that increasing attention is being paid to compounds contained in water, which manifest themselves in a certain labile state. Major changes in temperature and the influence of light and air can destroy these delicate formations within a short space of time. These formations, however, are what really matters. With regard to ordinary drinking water, this is particularly applicable to semi-bound carbonic acid and how it is incorporated in bicarbonates of salts. However, `uncombined` carbonic acid is also of great importance, since it is the essential contributing factor to the refreshing taste of good, high-grade springwater, and, as `associated` uncombined carbonic acid, it necessarily contributes towards maintaining the labile bicarbonates of salts in solution. Above a certain concentration the content of uncombined carbonic acid endows the water with aggressive properties and has a detrimental effect on metal surfaces, particularly with the presence of oxygen. The importance attached to the supply or exclusion of air is due to the fact that in groundwater, for example, pyrite does not decompose if air is excluded. The moment oxygen is introduced, as a result of human activity, sulphuric acid is formed from pyrite.

Attempts to transport certain medicinal waters while still retaining their properties have so far been unsuccessful. In those waters, whose efficacy is in part attributable to their content of certain unstable iron compounds, evidence of decay due to the entry of air and light can already be detected - although on first inspection everything appears to have been retained in the water qualitatively and quantitatively. A certain length of time after discharging from the mouth of the spring, all radioactive waters lose a great deal of their medicinal effect. Their emanational activity is greatest in very early stages, and when conducted in pipelines this can only be maintained with the implementation of very specific precautionary measures. Naturally, this is also valid for other types of water.

In Professor Dittler's opinion, if radioactive gas is added to medicinal water mechanically, it already loses half of its activity within four days. According to L. Winkler, the oxygen content of water lies between $6cm^3$ and

8cm³ per litre of water, depending upon the water temperature. This quantity is very slight compared to the amount of soluble carbon-dioxide in one litre of water, which decreases from 1500cm³ to 1000cm³ as the temperature rises from +4°C to +15°C. In general great care should be taken to ensure that the hydrogen ion concentration (pH) does not fall below 0.7 x 10 = pH7, since the aggressiveness of the oxygen will eventually damage the supply pipe. In addition the water's dissolved carbonates will also be precipitated as a result of the oxygen's activity.

Experiments carried out in order to determine the relation between water temperature and external effects of a purely mechanical nature have yielded no satisfactory result. Kerner tried to establish formulae demonstrating that the temperature t of a spring is a function of altitude and the petrographical composition of the mountain range. Thus, in the case of the springs at the foot of surface moraines in the Dolomites, for example, he presents the equation

$$t = 8.00 - 0.13h$$

according to which water temperature should decrease by about 1°C (1.8°F) with an increase in altitude in the order of 200-300m (600-900ft). However, J. Stiny states that the functional relationship between altitude and water temperature should not to be adhered to too rigidly, since many other factors are also involved, including the `motility of the air`.

Keilhack refers to heat influences active in the water itself, resulting from processes of oxidation and hydrate formation. Because of this function, these quantities of heat attain considerable importance. Where carbone appears in the concentrated form of hard pit-coal or brown coal, an additional heating effect occurs which is conditioned by the oxidation or combustion of coal seams in the interior of the Earth.

A phenomenon occurring in many places relates to the fact that springs deliver cooler water and rise higher in summer than in winter.[35]

In summer a positive temperature gradient exists from the stratosphere to the lithosphere. During this period cold, once strongly-oxygenated snow meltwater emerges into the light of day. In winter, on the other hand, there is a negative temperature gradient from the atmosphere to the lithosphere, and frozen ground prevents the infiltration of surface water so that the relatively oxygen-deficient water that infiltrated during the summer rises from the depths. In both cases the water has had an opportunity to charge itself with carbones over an extended period, and to restructure and ennoble itself appropriately under the influence of a favourable temperature gradient, with the result that such springs deliver excellent water. Both the length of time available for the water's processes of ennoblement inside the Earth,

---

[35] See "Temperature and the Movement of Water", latter part of footnote 44 - Ed.

and the oxygen content of the initial parent water are decisive. This is because if water richer in oxygen can reach deep levels, reconstituting processes take place in more lively fashion. Since snow meltwater sinking deeply into the cool layers of the ground has a greater content of oxygen than ordinary rainwater, it hence follows that cooler water appearing in summer must also be of a higher quality.

Many a hydraulic finding (today viewed from a purely mechanical standpoint, with scant consideration given to its physical aspect), will lead to an entirely different line of reasoning once physical factors referred to so far are taken into account. The result of this fundamentally different way of looking at things - from a *physical* instead of a merely mechanical point of view - is that my discoveries will never be incorporated into the contemporary complex of hydraulic opinion. My ideas will not be understood as long as adherence to the current one-sided approach persists. This applies to contemporary river-regulation as well, and especially to the internal destruction of water's character through its use as a raw material for machines. (The disastrous consequences caused by modern forestry are discussed more fully in *The Fertile Earth*, Vol. III of *Eco-Technology*). In recent years even chemistry has come to realise that it is totally inadequate to characterise a water or a medicinal water quantitatively or qualitatively by its given compounds of salt alone.

The restructuring processes constantly taking place in Nature can readily be emulated artificially in order to produce healthy, ripe water, once appropriate physical forms in which necessary restructuring processes can proceed can be manufactured.

The change in freezing and boiling points, evident in certain types of water, led to the finding that the freezing point of aqueous solutions is dependent on the number of molecules contained in one litre of water. Electro-chemistry was the first to take an approximately-correct path, in that it began to provide evidence of what really matters here. Whereas a solution of many organic substances (organic in the sense of modern chemistry) conducts very little electric current or none at all, precisely those substances (carbones) contained in various types of water are classified as electrolytes.

If the dissolving of carbone groups characteristic of water and medicinal water is properly carried out, ionisation can also occur without application of a low-voltage current. The fact that the conduction of electric current through aqueous solutions has been successfully achieved (in which ionisation of saline solutions occurs naturally without a detectable loss of electrical energy), provides proof of the above axiom. This phenomenon becomes all the more understandable and also gains in practical value when the explanation of the true nature of electricity is taken into account [*see The Fertile Earth*, Vol. III *of Eco-Technology*].

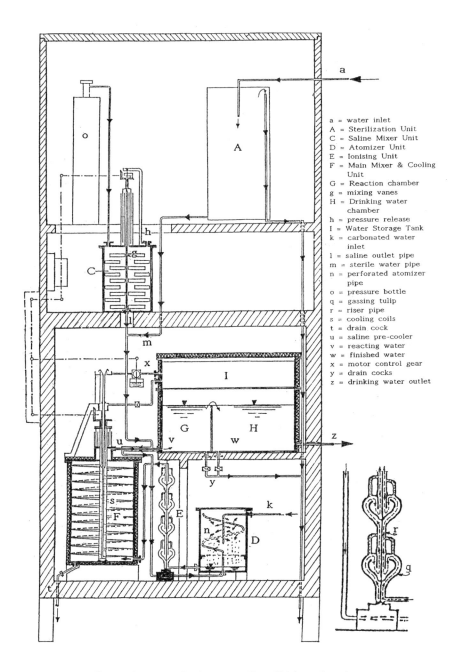

Fig. 12: Apparatus for the preparation of high-grade water.
(See Patent No. 142032 in Appendix)

While the method of describing water by specifying the salts in their dissociated form is certainly a small step forward, other energetic processes taking place in water are still far from being exhausted. The elucidation of these processes will radically change contemporary thinking and will enable practical applications of electricity, providing humanity with undreamt of possibilities for development.

The apparatus for the preparation and production of healthy drinking water (see fig. 12) cannot at the moment be described in more detail for patent reasons. Apart from this the means of producing other forms of energy directly from water, in a physico-mechanical way, should anyway be evident.

Had our scientists taken Nature as their teacher instead of consistently and stubbornly pursuing their own goals, we would doubtlessly have been spared our present misfortune. It is high time the many mistakes and errors made thus far (some of which occurred only recently while others have accrued from earlier epochs) should be rectified as quickly as possible in the interests of an increasingly destitute humanity. Any untoward delay in this necessary change in approach cannot be countenanced. To wait until the ponderous scientific establishment has laboriously adapted itself to new guidelines is out of the question.

Temperature and the Movement of Water

With other relevant texts on River Engineering

Between 1928 and the early 1930s, Viktor Schauberger was responsible for the building of nine successful - if hydraulically unconventional - log-flumes for the economic, damage-free transport of timber from inaccessible locations. As a result, the then Austrian government commissioned Professor Philipp Forchheimer, a recently retired and highly respected hydrologist of world renown, to study and report on Viktor Schauberger's theories and their practical implementation. Through their subsequent close association, not only did the two men gradually become firm friends, but Forchheimer became wholly convinced of the accuracy of Schauberger's theories and, in particular, the hitherto totally disregarded function of temperature as a vital factor in hydraulics and hydrology. So completely converted to Schauberger's pathfinding discoveries did Forchheimer become that he eventually confided to Viktor that he was glad to have retired, since he might otherwise humiliatingly have had to admit to his former students that most of what he had taught them was wrong.

This eventually resulted in the writing of Schauberger's first scientific treatise on temperature-related water movement entitled *Turbulence*. How this all came about is explained in the following article from *Implosion* Magazine (No.8) on Schauberger's visit to the Technical University for Agricultural Science. The substance of this treatise was based partly on Schauberger's centuries-old family connection with log rafting and, more recently, on his experiences with log flumes - in both of which water temperature played a crucially important role. Professor Wilhelm Exner, President of the Austrian Academy of Science ensured that this treatise was deposited under seal at the Academy on the 24th of January 1930, which was done under his personal supervision and witnessed by Corvette Captain Descowitz. The purpose was both to establish Viktor Schauberger's priority in intellectual property and to preserve it until such time as the hydrological world was ready to receive and to understand it.

As further confirmation of his change of heart, Forchheimer then requested Schauberger to write a complete scientific exposition of his theories under the general banner of *Temperature and the Movement of Water*, which was to encompass all aspects of water resources management, water treatment and supply, dam construction and hydroelectric power production. Using his powerful personal influence, Forchheimer then persuaded the editor of *Die Wasserwirtschaft*, the Austrian Journal of Hydrology, to publish them - a move that caused much opprobrium in scientific and hydrological circles. Since the readership of *Die Wasserwirtschaft* encompassed universities and institutes generally concerned with river regulation, hydrology and hydraulics, the various papers of Viktor Schauberger's that appeared were not aimed at the general public, but written specifically for the relevant specialists. This explains why they were couched in current river engineering and hydraulic terminology. Publication of these important treatises regrettably ceased upon Professor Forchheimer's death in 1931. - [Editor]

River Regulation - My Visit to the Technical University for Agricultural Science
[Extracts from an article from *Implosion* Magazine, 8, 1945.]

"If I am a fool then it is no misfortune, for then only one more fool will wander this Earth. Amongst the millions of mentally deranged it would barely be noticed. But what if I am not a fool, and that science itself has erred? Then the tragedy is incalculable!" (*Implosion* Magazine, No.51, p.29)

I did not visit the Technical University for Agricultural Science to learn. My purpose was to give a lecture on naturalesque river regulation to the teaching staff. It all happened like this. The world-famous hydrologist Professor Philipp Forchheimer was appointed by the then Ministry for Land and Forests to give his expert opinion on the log flume I designed and built at Neuberg in Steiermark. In the official commissioning report it was described as a technical wonder, because logs that were heavier than water (beech, larch, *etc*) swam in it like fish. In addition to this, the installation inexplicably achieved a 90% saving in running costs compared with usual systems of log rafting.

Professor Forchheimer studied the log flume for about six weeks without discovering how and why it refuted the hitherto supposedly irrefutable conformities with natural law. At the time I had no particular desire to divulge the secret. The professor, whose current outlook had received a nasty shock, therefore persuaded me to hold a question-and-answer session with the assembled body of teaching staff at the above Institute.

When I arrived I was taken up to the first floor into a large lecture room where about ten professors, including eminent hydraulic engineers, were foregathered. Professor Forchheimer first introduced me to the rector and then to the others with an inclusive gesture. I was led up to the president's chair at the head of the table and with a inscrutable smile and a wave of the hand I was invited to take a seat.

The general demeanour of the rector and the smirks on the faces of the other lecturers persuaded me to handle this worthy company in my own way.

The Rector opened the discussion with the words, *"Well, Wildmeister, would you be so kind as to inform us how we, as experts, should regulate our rivers naturalesquely (and here he laid particular emphasis on the word), so that as a result of these measures no erosion or water damage will ensue, but - and this you have already stated publicly - that advantageous after-effects only will be produced throughout the whole chain of evolution, as the ecological outcome of naturalesque methods of regulation."* After a moment's pause for reflection, I replied, *"Naturalesque river engineering is not easily explained in a few words."*

"Come now", interjected the Rector, *"perhaps you can highlight the essence of the matter with a few short phrases. Please keep it as brief and to the point as possible"*. To which I answered, *"In the same way that a boar passes water"*, wherein I stressed each individual word. The effect of this unexpected reply was as I had anticipated and desired - to shake them out of their complacency. At the end of a brief silence, and playing with his pencil, the rector addressed me very condescendingly. *"Wildmeister, we would be glad if you could choose your words more carefully and above all express yourself in more practical terms."*

At this juncture Professor Forchheimer stood up and declared, *"Your Excellency, gentlemen, I consider that Mr Schauberger's answer not only hits the nail on the head, but that it is also entirely accurate factually. Please follow me to the blackboard!"* Having arrived there, he completely covered the large blackboard from top to bottom with formulae totally unknown and incomprehensible to me. He did it with such gusto that he broke the chalk several times, throwing the pieces angrily aside. The rector looked on unwillingly and nervously drew himself together. Very soon the upper section of the blackboard was full, and with a great heave, Forchheimer shoved it up and half-filled the lower section with formulae.

Then he stood back and began to give a lecture. Naturally I was unable to understand a word he was saying. This then developed into a debate lasting some two hours, which was only terminated when an attendant appeared and reminded the rector of an appointment. The rector then excused himself quickly, shook my hand and said, *"We must discuss this matter again, but in greater detail"*. With this my first and last visit to the Technical University for Agricultural Science came to an end.

Professor Forchheimer took me by the arm and asked me to accompany him. He even forgot to take leave of his colleagues, to whom I quickly said good-bye, and swept impatiently out of the room. At the door he looked at his watch, was horrified to see how time had flown, and said quickly, *"Come to my house at nine o'clock tomorrow morning and then we can discuss this highly interesting matter in peace and quiet. I wish you had told me about this much earlier on, because it would be worth devoting a whole textbook to it"*. I nodded my agreement to his suggestion and then, adjusting his top-hat, he marched off.

The following morning I appeared punctually at 21 Peter Jordan Strasse.

Forchheimer immediately came to the point. *"Well now, let us discuss the curve you talked about yesterday in far greater detail. But before doing so, please explain how you arrived at the comparison with the boar. It is really very apt, but why this example?"* "Actually it doesn't come from me", I replied, "but from my father, who with these same words explained to his foresters how to arrange a log-rafting stream naturalesquely so that very heavy logs are able to float". Professor Forchheimer looked at me in surprise and then I began to explain the concept of `arranging` and its purpose.

"*The floatation of logs in the Klafferbach was an art requiring very special knowledge. Firstly there was only just enough water for the heavy logs, and secondly the bends were tight, which only a good `arrangement` was able to overcome. The purpose of this arrangement was to accelerate the passage of the logs with brake-curves.*" *"Just a moment!"* the professor interrupted, *"You mean that the water has to be braked in order to be accelerated? Not a bad idea, because in the process the water becomes compressed and its momentum increased."* "Not so, professor, that is not what I meant. The aim of these brake-curves is primarily to make the water rotate spirally about its own axis, like it does above every plug-hole." The professor scribbled something on his notepad, fiddled with his pencil and impatiently pressed me to continue.

"*Professor, have you ever watched a boar when it is urinating?*" He shook his head. *"Well then, try to imagine the shape of the curve produced by the flow of urine when the boar is running."* "That would be the most ideal cycloid space-curve, and it could not be constructed more beautifully", exclaimed the professor. After this observation, he tried to draw this peculiar curve, but quickly gave up. It is very difficult to draw, because the elevation is the same as the plan. He then tried to calculate it while I sat quietly and watched. He scratched first one ear and then the other, throwing away one sheet of paper after the next. He then declared that the curve would take years to calculate, even if the present state of mathematics was up to it.

"Professor", I replied, *"we are here concerned with a curve in which and through which life evolves"*. This led to a lengthy, wide-ranging, philosophical discussion which mainly revolved around the indefinable concept of `life`. Finally the professor said, *"I am a Jew and cannot agree. If you turn me upside down, all that will fall out are formulae. You think in a space that only you and no-one else knows - and therefore we cannot make any headway."*

On my way home I met the well-known author and former naval officer Captain Deskovic, to whom I related all that had happened. A few days later he called on me and invited me to visit His Excellency Professor Wilhelm Exner, who had a burning interest in my theories. Exner's welcome was very cordial, and he immediately enquired, *"Do you know anything about the brake-curves that maintain water's steady flow down steep gradients?"*. When I answered in the affirmative, he continued, *"Please understand me correctly. I do not mean any mechanical brake, but an inner safety brake".* He then called out to an elderly housekeeper for a cigar and requested that I explain my conception of this inner water-brake, because it was a problem that had tormented him for years.

"Before answering your question, your Excellency, I must first tell you how I view water and what I consider it to be." Exner smiled enigmatically, *"As you will"*, and invited me to continue. *"Were water actually what hydrologists deem it to be - a chemically-inert substance - then a long time ago there would already have been no water and no life on this Earth. I regard water as the blood of the Earth. Its internal process, while not identical to that of our blood, is nonetheless very similar. It is this process that gives water its movement. I would compare this inner motion, the origin of all possible physical movement, to that of a blossoming flower bud. As it unfolds, it creates a vortex-like crown of petals, in the centre and at the end of which stands the true secret of motion - life in statu nascendi, in the form of a concentration of movement.*

"I look upon this unfolding as the biological sequel to a preceding concentration of energy-matter (dynagen). It is the outcome of a form of radiation, which I view as a highly organised, vibratory process. Life itself, which springs forth as the final product of unfoldment from this ur-*fundament, is the highest conceivable concentration of dynagen. Through the agency of external environmental influences, and having manifested itself as a unique and unparalleled ostensible birth (because it is a precipitate), this concentration unfolds itself for the last time within a fraction of a second, only then to be extinguished. Whatever remains behind is a physical fruit comprised of raw materials in which inner levitative forces wane and which solidifies under the concentrating forms of heat generated by the incident light of the Sun.*

"In water the meanders continually bring about processes of concentration and unfolding. In a certain sense they represent `water-blossoms`, out of which growing radiation-emitting calyxes, or funnels, develop in an upstream direction. In their direct effect they are the cause of the braking of older (de-energised) river-water. The faster water drains down steep gradients - where it receives various impact-related impulses arising from encounters with resistances such as stones - the more powerful the reactive forces of recoil become. These, through the eruption (expansion) of the unfolding water-calyxes in an upstream direction, brake the apparently unrestrained flow on steep inclines as a result of these inner processes of growth.[36] *The secret of the outgrowth of new water, which takes place in the opposite direction to the flow, is concealed in the cycloid space-curve motion through which the water is made to pulsate."*

Professor Exner then told me that he had more or less understood what I had said, and that I should speak to no-one else about it. *"Please try to write*

---

[36] The lateral expansion of developing vortices restricts the space available for forward flow. - Ed.

down all that you have clearly stated, in simple language. I will then seal it, unread, in your presence and deposit it in the Academy of Science for use in the future. I am the chairman of the Academy, and I will see to it that your authorship rights are protected. I'll discuss this with Deskovic so that he can arrange for a cover note." This he did. I never saw Exner again because he died shortly thereafter.

Later on Professor Forchheimer informed me, *"I will be responsible for these expositions, and now you will be able to write about them in* Die Wasserwirtschaft. *However, please give me the manuscript to edit."* This is how my writings eventually came to be published - though later they were proscribed. Forchheimer then declared categorically that I should accompany him the following day to meet Professor Schocklitz and Professor Smorcek in Brünn. Professor Schocklitz showed us around his laboratory. He proudly showed us some glass plates across which water was flowing. This provoked the comment that I had never seen water flowing over such glass plates in Nature. As Professor Schocklitz took this remark greatly amiss, I diverted the conversation towards a turbine-blade lying in a corner, heavily pitted by the effects of cavitation.

Professor Schocklitz apparently had no idea of the decomposive after-effects of dynamitic substances,[37] which explained these cavitation-related phenomena in steel turbine-blades. According to the book *Deutsche Physik* these energy-precipitates, which range themselves in a particular direction, develop a peak performance of the order of 32,000 atmospheres and involve precisely the opposite products of synthesis. These I discussed in detail with Professor Smörcek, director of the university, immediately afterwards in his workshop - a 'workshop' is the only way such laboratories can be described, since they take no heed of the inner dynamic processes in water.

I drew Professor Smorcek's attention to the different effects of formative and destructive products of synthesis. Although he seemed very interested, no further contact with him eventuated. He said he would soon be returning to Vienna and would like to take me to meet Professor Schaffernak, director of the hydrology department at the university, in order to discuss this and other questions. Professor Forchheimer declined the invitation and explained that it would serve no purpose, since Schaffernak was too materialist. He could not explain, for example, why the waters of the Danube and the river Inn did not mix immediately at their confluence, but only much later and further downstream. This visit actually took place with Professor Smorcek present. I explained this phenomenon as arising from different conditions of tension and temperature which permit the waters to mix only after they themselves have first come to a state of common thermal and energetic equilibrium.

[37]*Dynamitic* substances: the violent, concentrated effect of oxygen in spacially-compressed, carbone-hungry form. - Ed..

Professor Schaffernak looked at me strangely and asked how I came upon such an idea. He asked some other questions too, through which I realised that he too viewed water as a chemically-inert substance, and its finely dispersed sedimentary matter, the bacteriophagous *threshold matter* in its most highly evolved state, as `impurities` in the blood of the Earth. In reality, these are the true sources from which negatively-charged fructigenic potencies are created. They intermix with descending, finely-dispersed and dosed seminal matter (oxygen). Through this interaction, water `comes to life`, and begins to pulsate.

A more comprehensive understanding of these processes was impossible for any of these scholars. It was only Professor Forchheimer who later murmured reflectively, *"I'm glad I'm already 75 years old. You cannot harm me anymore. The time will come, however, when you will be understood"*.

Over the course of the years I had many occasions to speak with scientists of world repute about the concept of `atomic disintegration and formation`. In Berlin, just before he was taken into custody, the famous physicist, natural scientist and Privy Councillor, Max Planck, was called in as an expert adviser during an interview I had with Herr Hitler. He only looked at me intently, but said nothing about my views. What he did say, however, was, *"Science has nothing to do with Nature"*. He then took me to the chief chemist at the Kaiser Wilhelm Institute. This conversation had disastrous consequences for me. I only mention it here merely to point out the dangers that arise if one discusses such matters with people who feel that their research has taken a wrong turn, but who want to maintain their own livelihood.

Naturalesque river regulation will only be understood after the inner process of motion outlined above has become common knowledge. For this a long period of re-education will perhaps become necessary. But first the tragic repercussions of today's purely mechanistic methods of river engineering will have to become starkly apparent through the occurrence of the most disastrous events. Everything we see around us, or are otherwise aware of, came into being through cycloid space-curve motion, which is the basis of self-renewal and evolution. It is a form of motion that can only arise when, owing to the braking resistances of the riverbank (the effect of meanders) the forward motion of the water is translated into a rotational one.

Shortly before his death, Forchheimer attributed cycloid brake-curves to me in his unfinished book. With this book, Forchheimer intended to refute everything he had implanted into millions of brains with his textbooks - he had taught that water had to be channelled as fast as possible by the steepest, straightest and shortest route from mountain to valley, so that it could flow into the sea having done as little harm as possible.

During this final period, Professor Forchheimer also realised that with this, he, as a leading hydrologist, had made the greatest teaching mistake of his life by destroying the naturalesque developmental pathway along

NATURALESQUE RIVER REGULATION

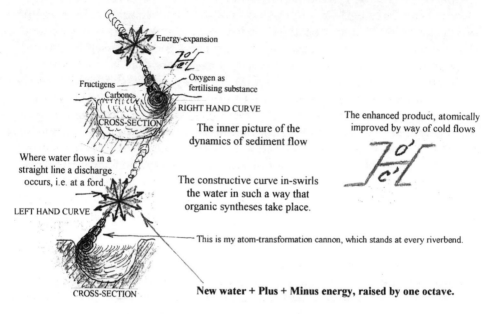

Water flowing constantly in a straight line, discharges its energy constantly.
Curves initiate a rejuvenation process, without which water dies.

Schauberger
Wien - Hadersdorf
Herzmanskystr. 1.

15./10. 1941

which, in a rhythmic sequence, the forward flow of water is intermixed with *formative* and transformative rotational motion. At the intersection of these alternating movements, under certain other specific preconditions, the formative, reproductive and upwardly-evolving power of *fructigenic matter* is released (see fig. 13 and refer to figs. 4 - 6 also). This becomes uni-polar, and so highly active that it is able to bind whatever the corraded river-gravel disperses into the water, or whatever the gills of the trout eject to promote their own mobility. In these life-renewing processes, whatever then radiates and expands upstream out of the calyx-like water-formations, as a surplus, conceals the riddle - *water*.

As a result of cooling influences, reinforced by naturalesque plantation of the riverbanks, negatively-charged fructigens become highly active. Finely distributed around the external surfaces of the water-calyx, the `aggressor` is to be found - the oxygen which becomes aggressive with warming and leads to a build-up of heat. It is only in this spacially-diminished (dispersed) and internally-passive state induced by cold influences, that the `aggressor` can be consumed (enveloped and bound) by what Goethe called the 'Eternally Female' and the 'All-Uplifting' (the carbones and fructigens).

However, if as a result of the elimination of cycloid space-curves in straight-channelled river regulations, the naturalesque evolutionary curve is lost and with it the possibility for self-renewal, then water becomes stale, insipid and sick. With this everything that owes its existence, its physical well-being and its increase in mental vigour to the water also disappears. Most importantly of all, it is quite futile to work towards an improvement in the present standard of living as long as the Blood of the Earth, the apparently ordinary-looking water that faithfully accompanies us from the cradle to the grave, is not given back its matriarchal rights. With these it can consume the `aggressor` in such a condition that self-renewal is not harmed - but which as the ultimate precipitate can only be put to good use. In this regard mechanical influence alone is insufficient. By means of a deterrent cold influence the inner destructive fury must first be removed from what has been dispersed, and this is done with the use of cycloid space-curve motion.

Turbulence – Concerning the Movement of Water and its Conformity with Natural Law
[Original treatise deposited under seal at the Austrian Academy of Science, 24th Jan. 1930]

The influence of water temperature has been dismissed as too insignificant for the purposes of stream management, and therefore for flood mitigation, timber floatation and rafting operations, water supply and dam construction in general, and also for the whole realm of hydroelectric technology. Documented variations in temperature yielded values arithmetically too small for any noteworthy effect on the results to be inferred.

It must be emphasised that the internal variations in water temperature are a *result of* the differences between the temperature of the water and the medium surrounding it.

If internal variations in water temperature are ignored, then the significance of differences between water and air or external temperature and therefore the *cause* of the *water cycle*, will likewise be negated. No word can truly express the vital role of the water cycle for all life on the Earth.

Of equal importance, if less obvious, are the *effects* of variations in temperature *within* the water itself, as will be shown later. Up to now such variations have been disregarded as immaterial for the purposes of all hydraulic calculation. Observations over many years, practical experiments and *correctly* carried-out measurements have proved that it is *absolutely imperative* to take internal variations in water temperature into account. *Their very exclusion -* elimination is out of the question - makes *all practical* use and exploitation of water *impossible. The understanding alone* of the important effect of these variations compels the *reappraisal and revision* of the *fundamental bases* of currently-held theories relating to the *whole sphere of river engineering*.

A *new*, hitherto neglected, but *extremely vital factor* is now added: the changes induced in the inner state of water through the stratification resulting from differences in temperature. Were this factor to be integrated into conventional theory, we would have to learn to *reformulate our ideas* about fundamental principles.

There is another omission in contemporary theory about the formation of many springs. Apart from commonly-known seepage springs where water above impervious strata is brought to the surface by gravity, there are also springs, lying far above any possible accumulation of water, which, breaking *all known laws*, surface rather like artesian wells much higher than the main water table. An example is a spring on the High Priel, which rises about 100m (330ft) below the summit, at an altitude of over 2,000m (6,500ft), and discharges water all year round.[38]

While it is not our purpose to explain the emergence of springs in general, the principles involved in these observations should be outlined briefly. In

---

[38] Mr Loew, a chief departmental engineer, also describes the following case: "In the Bukowina district a very cold, much sought-after spring, productive all year round, rose at the north-western base of one of the rock outcrops forming the summit of Mount Rareu. Since the immediate catchment area, considered to be the source of supply, amounted to only a few hectares, and both of the nearest higher peaks, the Dzamaleu and the Pictrossa, were separated from the spring by depressions lying far below it, the question remains open as to the origin of the spring's strong and constant flow. Mount Rareu is forested on all sides right up to the spring." Extract from a letter by Viktor Schauberger to the editor of Die Wasserwirtschaft, vol.7, 1931, p.106. - Ed.

the nature of things, as rainwater infiltrates the Earth's surface, it acquires the temperature of the strata through which it passes. Eventually it reaches a level with a temperature of +4°C (39.2°F). Naturally *this decisive stratum (the centre stratum)* is not horizontal, but roughly follows the configuration of the Earth's surface. The +4°C water impregnating *it has a specific density of 1*. Above and below the centre stratum the water gradually decreases in *density*. This water is therefore confined between two strata of different densities. Since their temperatures diverge from +4°C, they both endeavour to expand, thus exerting pressure on the centre stratum, which increases commensurately the greater the distance from this stratum. After the forces have reached equilibrium, the water in the centre stratum drains down the bed-gradient towards the lower side of the geological formation. The water strata surrounding the centre stratum, however, are *never* exposed *to the same pressure*.

Whichever of the two is exposed to *greater* pressure, it is forced back into the centre stratum. It *re-acclimatises* itself (to 4°C) and moves away downwards. According to the configuration of the ground surface, one or other of the enclosing strata soon comes under greater pressure and hence the centre stratum is fed alternately from above and below. This is how it is possible for water to emerge as a spring at such heights.

These high springs always exhibit temperatures very close to +4°C. The logical presumption is that the movement of spring water takes place within the centre stratum. The fact that water cannot be compressed at +4°C (39.2°F) may also play a certain role here - it must either yield to the pressure and make its way to the surface, or its temperature must adapt to the pressure. The latter occurs when the water in the centre stratum encounters an insurmountable obstacle. It is then absorbed into the neighbouring strata, and in this somewhat circuitous fashion it re-enters the centre-stratum by way of assimilation.

Enormous quantities of water are raised to immense heights hourly (through evaporation) and equally large volumes of subterranean water are forced upwards or expressed as springs on the highest mountain peaks. In both cases it is the hitherto ignored differences in temperature that continually disturb the state of equilibrium to such an extent. Small causes, too commonplace to be noticed, produce large effects.

To return to the actual theme - the effect of internal variations in water temperature on the movement of the water itself: It must be pointed out that these differences in water temperature would appear *totally to preclude* any state of rest in the water-body itself. Even in apparently motionless water very considerable movements occur - they are able to set large quantities of logs in motion. If an ostensibly still stretch of water is exposed to the Sun on one side only, then an inclined plane (thermocline) is formed through the warming of the water surface in the insolated area, which induces a flow towards the colder side and results in the formation of circulating currents. Therefore, even without a bed-gradient *movement of the water* takes place.

When water, comprising strata of different temperatures and therefore of different densities, flows down a riverbed gradient, these layers travel alongside and above each other for a long period without mixing. *The movement of every single particle of water down a given gradient is linked to a very particular velocity, which corresponds to its specific weight. If its specific weight is altered by the gradient (greater velocity, greater friction, increase in volume), then the water is unable to adjust itself readily to the new velocity without a transitional phase.*

The same thing happens when the specific weight is modified by *outside* influences, such as solar radiation. The water *breaks*, or in more common parlance, *becomes turbulent*, which is the *activation* of a hitherto-unrecognised *precision brake* in moving water, which operates with marvellous automaticity and which is normally actuated by the external temperature. The greater flow-velocity in cool weather and during the night suffices to change the water's volume and weight. The temperature of all the water filaments approaches +4°C, and hence density 1. As a result their specific weight ought to conform to the increase in velocity - in which case a constant increase in the rate of flow should occur. However, through the increase in flow velocity, the friction between water particles themselves and between water particles and channel surfaces will be intensified, resulting in a rise in temperature and a consequent increase in volume.

The picture thus emerges that:

- on the one hand an increase in flow-velocity occurs, and on the other a decrease in specific weight;

- the water filaments rupture and the water becomes 'turbulent'.

- the forward motion of the water will be resolved into the formation of vortices.

The greater the velocity of forward motion, the greater or more intense the *formation of vortices*. At a certain velocity this assumes such a violent nature *that water can actually be atomised in the water-body itself*, a phenomenon that manifests itself as a cloud-like formation.

In summary it can therefore be stated that turbulence is the *interruption of the forward motion* of flowing water. It occurs in the axis of flow (the position of greatest increase *in velocity*) in conformity with natural law, and arises due to the fact that in water *each and every specific weight corresponds to a particular velocity*. Turbulence therefore represents the *automatic activation of a compen-*

satory motion. It is the automatic and double-safe *brake* in all flowing water and in every channel.

Through knowledge of the spring and the way it comes into being, and with a clear understanding of the function of turbulence, every possible way to *make use* of water practically *in accordance with natural law,* and therefore *without limitation,* is made available to humanity.

0
000
00000
Deriving
from above
are the following
guiding principles and
basic propositions and with
————————— them the compelling necessity —————————
for the restructuring of
the whole body of
water resources
management.
00000000
00000
000
0

Everything flows, and
>   *all processes* in the *atmosphere*
>> are reflected in the interior of the Earth.

Guiding principles:

1. The body of water passing through a channel profile is *never* a homogeneous mass, but *always* exhibits strata of different temperatures.
2. In all channels the relation between flow-velocity and bed-gradient is primarily *dependent* on the thermal stratification of the water.
3. The channel profile affects the flow velocity to the extent that its form and composition exert an influence on the differences in the temperature of the individual water-strata.
4. The *profile* is a *product* of the processes that take place *within* the flowing body of water *itself.*

[Viktor Schauberger, Vienna, 1st January 1930]

"Temperature and the Movement of Water"
[An article by Viktor Schauberger published in *Die Wasserwirtschaft*, the Austrian Journal of Hydrology, Vol. 20, 1930]

To the Editor of Die Wasserwirtschaft:

Mr Viktor Schauberger has sent me the attached treatise concerning temperature and the motion of water. Since this has aroused my keen interest, due to the entirely new points of view it presents, which will not only be fruitful but pioneering in relation to dam construction and river regulation, I consider it to be in the public interest that this work be made known to a wider readership and the scientific world. With this in mind, I recommend the publication of this interesting article.

Yours faithfully, FORCHHEIMER , m.p.

Preamble

The increasing frequency of catastrophic floods in recent years, and the constantly increasing aridity in many areas, raise the question as to whether, in conjunction with other measures inaugurated by human hand, arbitrary systems of water resources management are not in part to blame for these evils. We are here concerned primarily with two factors, which need to be examined with this in mind: contemporary methods of river regulation and the increase in forest clearance.

Before addressing the theme itself, attention should be drawn to a very important factor hitherto ignored in all hydraulic engineering practices: t*he temperature of water in relation to soil and air temperature* as well as the *internal variations in temperature* (temperature gradients) *in flowing or standing water.* Since even small differences in temperature suffice to bring about obvious changes in the state of aggregation of water (solid, liquid and gaseous), it is quite easy to understand that larger variations in the internal temperature of flowing or standing water must have a decisive influence on its movement in and over the Earth. In the following section the hitherto-neglected interrelations between temperature and the movement of water will be addressed, and those errors will be identified which have arisen through disregard of this vital interaction.

Temperature Gradients - Full & Half Hydrological Cycles

The movement and distribution of water returning to the ground surface from the atmosphere is conditioned by the prevailing rainwater temperature and by the temperature of the surrounding air and ground strata.

If the temperature of the incident water is *higher* than the ground strata supposed to absorb it, then through cooling and becoming specifically heavier as a result, rainwater will readily be able to infiltrate the interior of the Earth. After having attained a temperature of +4°C (+39.2°F - also its condition of greatest density) and by sinking further, the water eventually arrives at strata of higher temperature, and by accommodating itself to these temperatures it becomes specifically lighter.

The further it sinks due to the pressure from the heavier water above, the greater its inherent resistance to downward movement, owing to its constantly reducing specific weight. Finally a state of equilibrium is established through which the all-important height of the groundwater table is regulated.[39] Under very particular conditions of pressure, a water stratum with a temperature of +4°C (the *centre stratum*) is formed within the general body of groundwater. In the case described above we are concerned with a *positive temperature gradient* , which is the rate of change per unit length between the temperature of the incident rainwater and that of the ground, expressed arithmetically.

This case also represents the *Full Cycle* of water, the full hydrological cycle. In reiteration of what has been stated earlier, this is characterised by the following phases:

- infiltration of water into the Earth;
- passage through the +4°C centre-stratum of the groundwater;
- purification at this temperature;[40]
- further sinking into subterranean aquifers due to its own weight;
- transition to a vaporous state due to strong geothermal influences;
- rising again towards the ground-surface with a simultaneous uptake of nutrients;
- cooling of the water and deposition of nutrients;
- draining away over the ground-surface;
- evaporating and forming clouds;
- falling again as rain, and so on.

In warm soils the +4°C groundwater stratum is missing. Hence the counterweight to the upward pressure from below is also absent. If the temperature of rainwater is lower than the uppermost ground-stratum, then the

---

[39] The weight of the +4° C water thrusts the lower, specifically lighter water downwards, which due to the rising temperatures thus encountered, becomes even less dense and hence develops an increasing resistance to further downward movement through its expansion against lower-lying resistances. - Ed.

[40] When sinking or rising again from the interior of the Earth, as described above, the water undergoes a reduction in temperature and to a greater or lesser extent precipitates its content of salts and other substances due to its diminished dissolving capacity, until at +4°C (39.2°F) it reaches the state where its capacity to hold salts and other substances in solution is at a minimum. This tendency to transude the dissolved matter is enhanced by the soil, which exerts a filtering action such that the water in the centre-stratum becomes increasingly pure - VS.

water initially sinks to a certain depth and there becomes warmed and specifically lighter. Finally it is forced up to the surface again by the pressure from below and, provided it does not evaporate immediately, drains away along the riverbed-gradient.

In this case we are concerned with a *negative temperature gradient* (water temperature lower than the surface-temperature of the ground). The full cycle no longer develops, but only a *Half Cycle,* namely precipitation of water earthwards, surface run-off, evaporation, cloud formation and re-precipitation as rain.

The following may throw more light on the temperature gradient and help in better understanding what is to be discussed later.

When an initially-negative temperature gradient (warm earth, cold rain) is coupled with a simultaneous drop in atmospheric temperature, the ground can be cooled to such a degree that the temperature gradient ceases to exist. The same thing can also happen with an initially-positive temperature gradient, if the infiltrating water is of sufficient quantity to warm the ground. In both cases, when a zero temperature gradient occurs, the drainage is conditioned by the actual riverbed-gradient until such time as the temperature gradient is reinstated through the action of friction and other factors. It is necessary in each instance to re-establish the required temperature gradient through the addition of water of the right temperature in order to brake the water's free and almost resistance-less flow down the inclined plane of a riverbed.

The Groundwater Table

The height of the groundwater table fluctuates according to the temperature of the ground strata, which are also affected by the temperature of infiltrating water-masses. Air temperature also plays a major role.

Where localised impoundment of water-masses occurs, cold bottom-water influences the temperature of surrounding ground strata. These are cooled, and in this way a stable, positive temperature gradient is created, since in this instance rainwater will always be warmer than the colder ground. These are the preconditions for the infiltration of rainwater. As a result, the groundwater table will not only be raised, but the absorptive capacity of the soil will also be increased laterally and vertically.

The previously-described +4°C centre stratum in the groundwater will be displaced downwards owing to increased pressure from over-lying water-masses, thus overcoming the resistance to further downward penetration of warmer and specifically-lighter water lying below the centre stratum. This leads to the formation of a natural subterranean reservoir, a retention basin, which inhibits rapid surface drainage and gives rise to the full hydrological

The Groundwater Table

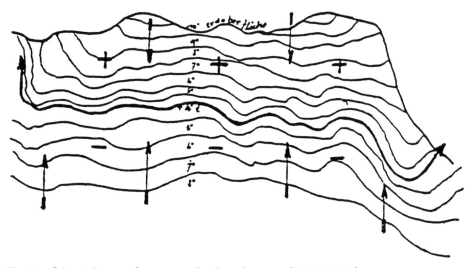

Fig. 14a: Original diagram of a cross-section through a groundwater reservoir

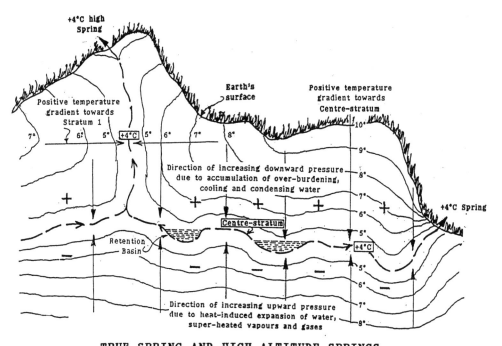

Fig. 14b: Amended diagram of a cross-section through a groundwater reservoir

cycle. The release of water from this reservoir then follows. As will be described later, the lateral expansion of the centre stratum (formation of springs) can also occur due to pressure acting on it from above and below. Fig. 14a[41] below, schematically illustrates a cross-section through a groundwater reservoir. This depicts how not only the lateral but also the upward flow of the centre stratum (under the greatest pressure from all other strata) can come about.

In contrast, as described previously, high surface temperatures permit hardly any water to infiltrate into the ground at all. The accretion of groundwater ceases, or it only accumulates in small quantities at great depths. Through evaporation of small residues of groundwater still present, the ground becomes increasingly incapable of absorption. Due to the effect of too high a ground temperature, percolating water will develop into surface run-off, leading to rapid re-evaporation connected with it (a half-cycle). In this case no accretion of groundwater in the previously-described sense therefore comes about. In such districts hot springs frequently make their appearance, forced up to the surface through fissures by upward pressure from below, for which no counter-pressure exists due to the lack of over-burdening groundwater.[42]

The absorptive capacity of the ground is thus dependent on the conditions of temperature which give rise to the regulation of the groundwater table as described above, and hence on the existence and height of the groundwater-table.

In summary it can be stated: a positive temperature gradient is the precondition for the soil's ability to absorb, for the accretion of groundwater and for the creation of the full hydrological cycles associated with it. A negative temperature gradient prevents the accretion of any groundwater and gives rise to half-cycles only.

[41] It is known that Viktor Schauberger was unhappy with the original diagram and the translator has submitted his own interpretation in figure 14b. - Ed.

[42] "In the years 1911-1914 and 1918-1924, I was able to observe that on the occasion of sharply contrasting temperatures - for example, with the onset of hard frosts after a very dry autumn - and after the streams had dried up, hot water was forced upwards from great depths through craters, fissures and holes. It is characteristic of this case that grotto-olms, (a species of salamander) about 30cm long, will be thrown up at the same time, causing a large influx of crows to the area. A particularly striking case happened in the presence of witnesses in October 1923, in which year there was a dry autumn. The 'E'-impoundment of the Steyrling timber-flotation system was suddenly filled with hot water under an outside temperature of -12°C; within 10 minutes about 1,500m3 (1,900 cu yds) of water welled up out of the old crater at Pärnstall-Fütterung, filling the basin. After cooling off, these waters subside and, forming whirlpools, and with a great rumbling commotion, fall back into the Earth again. The districts are: Ödsee (Parish of Grünau), locally known as the 'Roaring Lake', Lahner Alpe (Stier-Kar), Bauernschlag Alpe (above the new hunting lodge), Kasbergmulden Alpe, Luketerwald, Mangstlberg (from the forest house 'Hofjäger' to Pärnstall, especially in the vicinity of the so-called 'E'-impoundment. Conversely, after protracted periods of rainfall and with the advent of warmer weather, cold water-masses will be driven upwards. After having cooled and with much turmoil they disappeared back into the depths within a short space of time. I would like to point out that these areas are large enclosed hunting reserves and are not generally accessible." This is an extract from a letter to Die Wasserwirtschaft, Vol. 3, 1931, p.47. - Ed.

The Drainage of Water

The drainage of water below the ground surface (groundwater-flow) occurs as a result of pressures exerted from above and below in conjunction with temperature variations obtaining in the groundwater and the surrounding ground strata. What alone has been construed as groundwater flow up to now was merely the drainage of groundwater overlying an impervious layer - drainage which moves down its inclined surface until it is re-integrated into the full cycle. Water is able to discharge over the ground surface under two conditions:

1. With a negative temperature gradient (as described in *Temperature Gradients - Full and Half Hydrological Cycles*) this occurs immediately!
2. With a positive temperature gradient it takes place only after saturation has occurred, and the groundwater table rises towards the surface under the influence of the Sun's heat.[43]

This also explains a phenomenon often observed in the mountains: rainfall over several days causes no appreciable increase in the flow of water in the associated receiving streams. The rainwater is almost completely absorbed by the ground. Only after the onset of warmer weather does a flood-discharge enter receiving streams. Cold groundwater rises to the ground surface, which by this time has been warmed by the Sun and warm air. The earlier positive temperature gradient is transformed into a negative one; the water flows away. Country folk say "the mountain is pissing".[44]

In Case 1 above, the preconditions for the creation of floods are far greater and all the more so, if as a result of the direct run-off of water resulting from an initially negative temperature gradient in the ground (cooler higher stra-

---

[43]The Sun warms up air-strata lying above the Earth's surface. these expand, and becoming specifically lighter they stream upwards. The air pressure falls, and the state of equilibrium is disturbed. The pressure acting on the groundwater from below lifts it up until a new state of equilibrium can be established. The effect of the pressure from beneath is reinforced by the sucking effect of the rising air. This also explains why the groundwater table rises by day and falls by night, and why springs flow faster at night than by day - for with the general sinking of the groundwater at night, additional pressure is exerted on the +4°C centre-stratum, causing more water to be expressed either vertically or laterally. In just the same way, retention of groundwater on steep slopes is also to be attributed to the equilibrium-related phenomena outlined above. - VS.

[44]"To give a more detailed example: in the summer of 1920 it once rained incessantly for almost a week. At midday on Saturday it was bright and warm again. In spite of so much rain, no rise in the streams could be detected. At 4 o'clock on this Saturday afternoon, the author's 6 year-old son passed by the so-called 'Walls' in full sunlight on his way home from school. Towards evening the author also returned home in company with a man from the hunt. For no apparent reason a flash-flood suddenly burst forth, a rapid ascent of the rock-walls providing the only means of escape. This case is clearly fixed in the author's memory, because of the danger to which the child had been exposed. The reason why under certain preconditions flooding must occur after rainfall ceases were discovered only many years later. In this regard chief departmental engineer

ta, warmer lower strata), a positive temperature gradient is developed (effect of friction), when the water tends to infiltrate the ground gradually, loosening and carrying away boulders and pebbles. A thermal surface-gradient is now added to the physical riverbed-gradient. A further increase in run-off velocity and the power to shift pebbles, gravel and sediment ensues. Once the present positive temperature gradient again becomes negative, bends in the river are formed in the lower reaches through turbulence, and thus a mechanical deceleration in the rate of flow occurs. Suspended sediment is deposited and the oncoming water-masses become backed up. The result: flooding.

In Case 2, if saturation of the groundwater basin occurs as a result of a stable positive temperature gradient, then groundwater (springwater) that now surfaces is colder than the ground strata lying directly beneath it. The temperature gradient has been reversed and has become negative again. Rapid drainage of the heavy water-masses follows. As a result of relatively low temperatures in the ground, cold, heavy, excess water from the Earth's interior now drains off, achieving a positive condition only gradually, because the specifically-heavier water warms up very slowly. Since in the upper third of the catchment area the slope of the riverbed is usually extremely steep, turbulence is created, and hence bends in the more horizontal parts of the river are formed.

The further transition from a negative to a positive temperature gradient therefore takes place very slowly, and the incidence of strong turbulence again leads to excessively sharp horizontal bends and to the deposition of boulders, pebbles and sediment, the gouging of pot-holes and the dislocation of the channel bed through mechanical action. The immediate result of this type of discharge is a widening of the channel, a heaping up of broad banks of boulder-gravel, and evaporation or subsidence of water in the churned-up riverbed. In this process the riverbed has again been exposed to the influence of external temperature (already typical of alpine flow-regimes, and always associated with an asymmetrical profile - a deepened

---

Loew describes the following case: 'It was in July 1902 or 1903, the exact date and hour escapes me. The place: Bukowina, the catchment area of the Moldawitza, a large enclosed area of primeval forest with typical primeval forest cover. After a prolonged period of dry weather a general and persistent rainfall of varying intensity occurred, lasting about six days. I remember very distinctly that on the last day of rain general flooding was expected. It did not happen on that day, however, but only after the rain had stopped - and when it actually did flood, it did so with unparalleled force, as though the forest had suddenly lost its ability to retain water.' Mr Thaler, the Minister of Agriculture, also kindly made the following example available: 'My house lies high up on a mountain, and naturally water is of major importance. As happens every year the so-called 'May-water' appeared in 1911 just next to my house. 1911 was a conspicuously dry year. With much anxiety I kept a constant watch on the little streamlet. With increasing dryness, the spring suddenly began to discharge about 20 paces higher up the mountain. The drier the season, the more abundant the water from this new, higher-lying spring, and the warmer the temperature and the hotter the summer, the colder the water was." From a letter by Viktor Schauberger to the editor of Die Wasserwirtschaft, Vol 7, 1931, p.106. - Ed.

bed on one side and a gravel shoal on the other). The discharge of water takes place immediately in exactly the same way as in case 1. The rupture of the riverbank is the result. In times of flood mechanical braking is effected by the bends thus formed, and banks are breached even further. The situation is even worse than before and again the result is flooding. In order to avert the danger of flooding completely it is necessary to *eliminate* both extreme cases 1 & 2 *artificially*.

By building dams incorporating appropriate provisions, thermal conditions and rate of discharge can once again be regulated - where these have been altered inappropriately owing to the shift in the temperatures and the associated temperature gradient of the ground strata. The discharge conditions of these regulating dams can be automatically adjusted thermally and quantitatively to the prevailing daytime temperature. In this way both of these extreme cases will be avoided and will also be modified automatically so as to fall within the intermediate temperatures of the discharge. With increasingly finer adjustment of the simple apparatus proposed for these dams, temperature gradients suited to the mean seasonal temperature are progressively developed in the river, and in this way it is possible to reduce the danger of flooding at its inception and gradually to avert it.

There is no danger of flooding because, by adjusting the temperature of the discharge to the mean annual temperature, the correct ground temperature gradient can be re-established. This results in the restoration of the absorptive capacity of the ground, the proper regulation of the groundwater table and with this, the formation of the vitally important *retention basin*. Through the correct adjustment of the discharge conditions, an orderly further drainage of water over the ground surface results. *No localised evaporation* takes place, and because of this no rapid succession of rainfall occurs, restricted to a limited area. In other words, the well-ordered conditions of the full cycle are re-established.

Where de-watering or drainage of the ground is desired, it is likewise possible to make unwanted, stagnant water disappear by creating a temperature gradient (positive temperature gradient) conducive to this situation. It is therefore possible to produce a full cycle or half-cycle at will. However, dams that have been constructed so far have produced half-cycles only.

In this connection the meaning of full and half-cycles should again be clarified. The *full hydrological cycle* involves the entry of water into the interior of the Earth, the creation of the necessary groundwater body, the detention of run-off water and by this means to forestall or reduce the danger of flooding. Cold springs are also continuously formed, whose waters reduce the temperature of receiving streams and help to inhibit over-rapid evaporation downstream.

With the *half-hydrological cycle,* by comparison, a familiar condition occurs where rising water vapour is produced almost uninterruptedly. In other words, a continuous contribution is made to the mass of atmospheric water and the recurrent precipitation associated with it. One flood therefore gives rise to the next.

Basic Principles of River Regulation

It is vitally important to achieve the proper conditions of discharge not only in the above sense, but also in the regulation of waterways and the formation of their banks. The aim of contemporary river-regulation practice is to effect the fastest possible drainage of water, through bank-rectification and bank-stabilisation with artificial structures. This type of regulation, however, is thoroughly one-sided and does not fulfil its purpose.

It *cannot* and *should not* be the task of the river-engineer to correct Nature. Rather, in all watercourses requiring regulation, his job should be to investigate Nature's processes and to emulate Nature's examples of healthy streams. Here again the most crucial factor is the interrelation between water and air temperature, which cannot be disregarded in any regulation.

The natural regulators of the drainage of water are forests and lakes. By cooling the ground in their immediate vicinity, forests create a permanent positive temperature gradient, resulting in the formation of groundwater reservoirs which have a delaying effect on rainwater discharge. Once again, cold springs issuing from these groundwater reservoirs quickly enter receiving streams, cool the main body of water and thereby inhibit premature evaporation as the water flows along the channel. The cyclical movement of water - the transfer of water from the ground to the atmosphere - will be slowed down and distributed spacially along the length of the watercourse. These cycles do not take place over relatively small areas, so that *one* fall of rain or *one* flood does not necessarily give rise to the next.

Where forests have been felled and natural lakes are absent, it is necessary to create a substitute: an artificial impoundment of water, which *must be correctly built and properly operated*. Only then can it bring about the specified functions of groundwater-recharge, detention of run-off and the creation of a proper temperature gradient.

Indeed, impounded lakes are often built to enable the orderly management of water resources. However, these have not always proved satisfactory and have often achieved the opposite of what was desired. To be more specific: as constructed and operated today, impounded lakes are nothing more than storers of water. They collect the water and fulfil the function of rainwater detention, but almost always produce a half-cycle. The water remains on the ground surface (no infiltration) or evaporates soon after its release.

Precipitation in the vicinity of existing reservoirs becomes irregular and increases or decreases according to the orientation (wind direction) of the valley. The normal flow of water in the middle and lower reaches diminishes, the groundwater table also sinks in the middle and lower reaches for the same reason and the productivity of the soil in these areas noticeably declines. This happens for the sole reason that a thoroughly one-sided temperature gradient is created in the downstream flow-regime because of the way reservoirs are constructed and through the continuous release of either specifically-heavy or light water, depending on whether water is released directly from the bottom of the reservoir or via a spillway from the top. Both types of discharge lead to the extreme cases outlined earlier and thus to the generation of half-cycles, with their well-known detrimental effect in the spawning of floods and the resulting damage.

It is therefore the purpose of a properly-constructed reservoir, equipped with the requisite discharge-control systems and starting at the dam itself, to regulate the temperature gradient of watercourses continuously in such a way that these depredations can be avoided with certainty. With this method of regulation of the temperature gradient, expensive but usually inadequate installations in the channel itself become unnecessary.

Correctly-constructed reservoirs, as such, are those in which the movement of the water, though slight, will be enhanced by the development of a strong temperature gradient. Thus, by means of the proposed equipment, cool water strata will continuously and automatically reach the water surface, significantly reducing excessive evaporation - with its unwelcome consequences - which has occurred over these reservoirs up to now.

Dangers of flooding can only be prevented in a practical way if, with the use of naturalesque methods of control, water is not returned to the atmosphere as rapidly as possible - as has hitherto been the case, but is able to fulfil its true function. This is the establishment of the full cycle in its roundabout route through the Earth, and with it the supply of nutrients to the soil. It is evident that to date two cardinal errors have been committed: by draining water too rapidly *over the ground surface,* it is returned to the atmosphere too quickly, thereby causing renewed precipitation and flooding. More importantly, the water was thus robbed of its most important purpose of infiltrating into the ground. By inhibiting the full cycle, the supply of nutrients to the soil was also cut off.

River engineering carried out without consideration of the temperature gradient, and concerned exclusively with drainage of the water-masses down the riverbed-gradient, ultimately leads to disturbance of the proper sequence of temperature gradients, or to development of a one-sided temperature gradient - and hence to catastrophes and inundations. In France, for example, these must now occur with increasing intensity.[45] Moreover they will also become common further south until the current misguided practices cease.

The Interrelationship between Groundwater & Agriculture

In the preceding section attention was drawn to the mistakes that have been made in the execution of hydraulic engineering projects and indications were given as to how they can be avoided. In the following, the devastating consequences that ensue from the incorrect management of water resources are to be given special emphasis.

Through mismanagement of waterways, not only are riverside communities exposed to a direct and acute threat but, what is far worse, they are also threatened by an insidious evil, a reduction in soil productivity. This manifests itself in the retreat of groundwater or its other extreme, swamp development. If we note the changes that have occurred in areas under food production within the space of a single generation, and if we consider that today (1930) in Austria hardly any grass grows where once our grandfathers enjoyed rich farmland, it is clear to us how fast the productivity of the soil is declining. For example, the areas under wheat and rye cultivation have fallen from 273 million hectares to 246 million over the last 30 years.

This decline in yield is particularly marked in mountainous regions, which naturally are the first to feel the full force of the retreat of groundwater. On alpine pastures, where previously the raising of 100 head or so of cattle was of no consequence, today those with grazing rights squabble over the fodder required for a single beast. The previously almost inexhaustible, pastureland is today insufficient even for a fraction of its former carrying capacity.

The reason for this decline in soil fertility is purely and simply that the groundwater table has subsided and is continuing to sink further. The soil, which ought to produce a good yield, must be replenished constantly with additional ingredients required by the plants for growth. The carrier and distributor of these substances is the groundwater, which in its internal cycle constantly brings up fresh nutrient salts from the interior of the Earth.

---

[45] The same applies to rivers as applies to human beings. If they are sick and their functions are impaired, then external intervention and confinement are insufficient to effect a fundamental cure. Healing and health essentially lie in the blood of human beings and in the water of rivers. These need to regain their original purity, coolness and energy. They will then be able to restore health to their environment - the body - the channel and riverbanks. On the other hand, if we merely patch things up externally, if we encase these vessels - arteries - in concrete and steel, if we enshroud the river in a strait-jacket of walls and embankments, then we turn it into a defiant rebel. It will become our enemy, when all it actually desires to be, and could be, is our friend. Force, destruction, war between human and Nature are the unavoidable result. - Werner Zimmermann - *Tau* 137, page 8.

If the groundwater recedes, then the natural supply of nutrients ceases. Artificial fertilisation and redoubled effort constitute only a temporary and incomplete substitute for the natural supply of material. Atmospheric precipitation only moistens the ground and contains no nutrients for the plants. Nature herself is not responsible for the constant increase in the dessication of the Earth's surface caused by the sinking groundwater table. Rather, since time immemorial, it has been the unconscious hand of humans that is to blame for the constant lowering of the water table, and with it the withdrawal of natural nutrients.

The reason why water has been generally mistreated is because the importance of the temperature gradient for the movement of water according to inner law has been unknown until now. In consequence water was generally mistreated. In exploiting water's inherent energy for electricity generation, for example, arbitrary structures have been installed in channels which in many cases have affected the water destructively. Attempts have been made to regulate rivers by their banks, naturally producing negative results. No thought was ever given to the re-establishment by other means of the river's equilibrium, which was disturbed by structures in the river itself and through forest clearing. The method referred to here - artificial re-establishment where necessary of temperature gradients that under normal circumstances come into existence naturally - is the only correct solution to the problem of bringing about natural drainage of water or its retention in the ground. Only by pursuing this course or by making use of these findings can further subsidence of groundwater be prevented, and a further drop in soil fertility avoided. Only in this way will it also be possible to avert the devastation of floods, and to transform water once more into what it always was and always must be: the *Giver of Life.*

Fundamental Principles of River Regulation
– with due Regard to the Status of Temperature in Flowing Water

[An article by Viktor Schauberger published in *Die Wasserwirtschaft*, the Austrian Journal of Hydrology, Vol. 24. 1930, pp. 498-502.]

The most important factors affecting a waterway will now be addressed in broad outline and the techniques will be presented for regulating waterways in ways that correspond to Nature's laws. Questions of detail will not be dealt with here.

Turbulence Phenomena in Flowing Water

When an ideal liquid flows down an inclined plane without friction, individual filaments of the current ought to move along parallel to each other. Moreover, according to the law of gravity, this motion ought to accelerate uniformly. This never happens in Nature, however, since friction occurs between liquid and channel surfaces and between particles of the liquid themselves. As energy is dissipated in this process, motion is no longer accelerated, but is uniform - if pulsations and other irregularities are discounted.

In the case of a non-ideal, viscous liquid, as long as water-movement is stratified *(laminar)* - surface friction for the moment excluded - a certain amount of energy is transformed into heat. At a particular velocity, which varies according to water temperature, laminar motion transfers into a *vorticose*, turbulent one. With turbulent motion a certain amount of energy is also converted into heat, as was demonstrated by Barnes' and Coker's experiments,[46] and a further amount of energy is dissipated through exchange of momenta. In this context Forchheimer states that[47] *"in vortical motion the more central flow is not only transformed into heat but also into vortices, and conversely an acceleration of the more central motion can also possibly occur through a reduction in vortical activity, although no experimental evidence for this is available"*. The author's own observations reveal that:

- turbulence is at a minimum at a water temperature of +4°C (+39.2°F) under equal conditions and in identical profiles;

[46]See Forchheimer, *Hydraulik*, 1914, page 51.
[47]See Forchheimer, *Hydraulik*, 1914, page 27.

- turbulence and the associated decrease in velocity become more pronounced the more the water temperature diverges from +4°C;
- it is possible to achieve an acceleration in the central flow by inducing a decrease in water temperature towards +4°C.

Fig. 1 shows the exceptionally strong occurrence of turbulence and vortices where a hot spring flows into the Tepl near Karlsbad (Karlovy Vary). If the hot spring water is blocked off temporarily, the water in the Tepl flows downstream with considerable velocity due to the pronounced slope of the stream-bed at this spot. After re-introduction of hot spring water, this is reduced immediately to an extraordinary degree.

The enormous effect of water temperature on turbulence and velocity can also be observed at a log-flume in Neuberg (Steiermark). Here in a half-round, 2km (1.2 miles) long wooden flume, measurements of temperature and velocity were made during the floatation of timber. In the morning when the water temperature was roughly 9°-10°C (48.2°F - 50°F), a block of wood required about 29 minutes to cover the distance. At midday, with a water temperature of 13°-15°C (55.4°F - 59°F) and under otherwise equal conditions, it took 40 minutes.

A further example of this concerns water supply to the turbines of a board mill in North Austria. The water supply consists of two 2km long concrete channels. One draws its water from the so-called *Cold Murz*, the other from the warmer *Still Murz*. The former flows towards their common intake along the shaded side of the valley, the latter on the sunny side. With the canal profile at full capacity the normal flow of water from the *Still Murz* amounts to about 860 litres/sec (189gals/sec). According to the observations of Mr Brückner, the factory director, and Mr Patta, the works manager, on occasions when the water temperature of the *Still Murz* approaches that of the *Cold Murz*, and the temperature gradient in the supply canal from the *Still Murz* becomes positive, under certain circumstances (such as at night) the volume of water increases to 1,800 litres/sec (396 gals/sec). Despite the constriction of the intakes above the turbines, the output of the turbines increases, resulting in an increase in power generation equivalent to the thermal output of one wagonload of coal per night.

Temperature Gradient, Riverbed Slope and River Bend Formation

The formulae applied today to the calculation of flow-velocity in channels encompass geometrical profile of the channel, roughness of channel wall-surfaces and gradient (riverbed-gradient, slope of the water surface or energy lines). What these formulae do not take into account are the physical properties of water, such as viscosity and specific weight, which vary with temperature. However, it is important to take note of the temperature regime in the direction of flow - the temperature gradient or rate of change in temperature per unit length in the direction of the downstream flow. The temperature gradient is described as positive when the water temperature approaches +4°C in the direction of flow, and in the opposite case, as negative.

If for example the temperature at point A of a channel is $t_1°$, at a lower point B is $t_2°$, and if $t_1 > t_2$ *(positive temperature gradient)*, then along this stretch an increase in velocity occurs due to a reduction in turbulence. Horizontal transverse vortex-trains and turbulent formations become smaller. In the opposite case, where $t_1 > t_2$ *(negative temperature gradient)*, the incidence of turbulence increases owing to a rise in temperature and an ensuing loss in kinetic energy, which expresses itself as a decrease in velocity. The tractive force becomes less and deposition of transported sediment follows.

In the section relating to tractive force and the movement of sediment, Robert Weyrauch states in his book, *Hydraulic Calculation:* [48]

"*S_0 [boundary shear force[49]] is dependent on the provenance of the sediment, and is therefore constant for a relatively short stretch of river without the presence of affluent streams. In the case of longer stretches without affluent streams it diminishes in a downstream direction.*"

In the above example the reason for this is obvious - a case of negative temperature gradient. Where secondary streams exist (which reintroduce colder water into the main stream and thus usually effect an increase in flow-velocity through a reduction in turbulence), weakening of the *tractive force* does not occur. Tractive force is maintained or increases with a positive temperature gradient and decreases with a negative temperature gradient.

This phenomenon becomes all the more important when studying changes in the riverbed. Assuming a uniform discharge of water, the bed-gradient remains constant, or will become greater with a positive temperature gradient and smaller with a negative temperature gradient. Where the volume of water increases in conjunction with a negative temperature gradient, the morphology of the riverbed itself is not substantially altered, whereas under these conditions ruptures of the bank do occur as the central axis of the current oscillates from one side to the other. With an increase in the volume of water and a positive temperature gradient, the riverbed will be attacked and deepened. The watercourse straightens out and river bends previously formed through deposition of sediment will be evened out. Under certain circumstances, with a sudden drop in temperature and

---

[48]*Hydraulisches Rechnen*, 4th edition, page 68. - VS
[49]See footnote 3. - Ed.

atmospheric pressure (such as clear skies after a flood), especially at night, the descending flow of water can become even more dangerous than quantitatively greater masses of water under a negative temperature gradient in warm, rainy weather conditions.

The mean central riverbed gradient which develops over the course of time is affected by the mean annual discharge and the temperature gradient corresponding to the mean annual temperature, wherein the mean annual temperature and the amount of rainfall are to a certain extent interrelated. In those years where larger fluctuations in the mean annual temperature occur, there will also be relatively greater changes to the riverbed.

Measurement of temperatures in the same river cross-section indicates that temperatures vary according to location. Also, during the course of a day the place of the greatest flow-velocity (flow-axis, central core of the current) also changes its position within the profile *laterally* as well as vertically. While the lowest temperature is always to be found in the central core of the current, it increases to a greater or lesser extent towards the periphery. During the day the line of the central axis of flow lies closer to the shaded bank, since that is where the heavy water accumulates, whereas the lighter water flows along the sunny side. At night, due to the enlargement of the heavy water side, the current core migrates towards the centre of the channel. With a negative temperature gradient, the current core lies close to the water surface, and with a positive temperature gradient, deeper down.

During the floatation of timber the following phenomenon can be observed: if the temperature of the surroundings is lower than the water temperature (temperature gradient decidedly positive - water cools during flow), floatation takes place with the greatest of ease. The logs stay in the middle of the channel and float down the clearly-defined central axis of the current. On warmer days, especially towards midday, timber becomes stranded. Log-jams happen easily, because the flow axis wanders about (transverse currents due to turbulence) and does not keep to a centralised course for a prolonged period, as it does with a positive temperature gradient.

In section *1-1¹*, in the stretch of river shown in fig. 15, the axis of the current still lies in the middle of the river. If the values of the mean flow-velocities in each vertical of the river profile are plotted vertically, and an energy-line is drawn, then as is to be expected, the energy falls off to a greater or lesser extent towards the river's edge. If this decrease exceeds a certain limit, then it is obvious that this condition can only be unstable and even small causes will suffice to alter the *status quo*.

If, for example, the bank at **1** is shaded[50] and the bank at **1¹** exposed to the Sun, then at **1¹** the water will be warmed, becoming specifically lighter, and

---

[50] The initial cause can also be purely mechanical, *e.g.* one bank is rough, the other smooth. - VS.

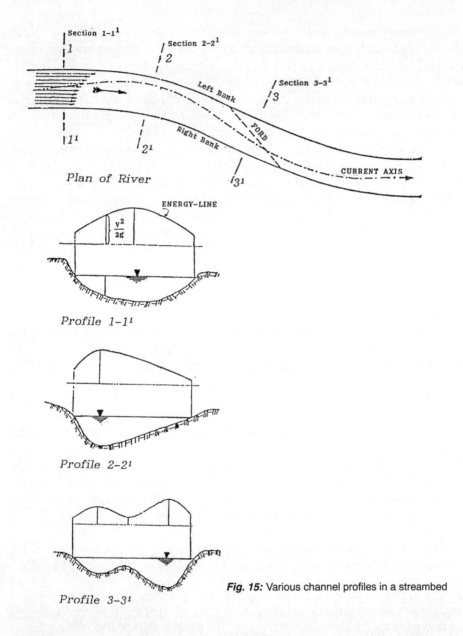

Fig. 15: Various channel profiles in a streambed

due to increased turbulence will flow more slowly here than at *1*. As a result of this, heavy water flowing along the left bank will advance more rapidly, already initiating the first beginnings of circular motion, shown in fig. 16.

In this instance the point of rotation lies beyond the profile of the river. A new condition of equilibrium is established (profile *2-2¹*). This circular

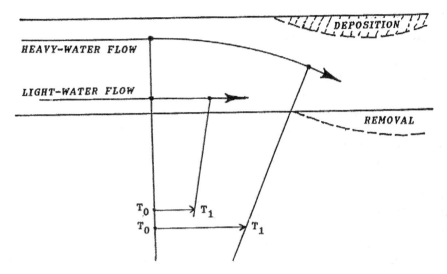

Fig. 16: The formation of river bends

motion continues until the respective temperatures and velocities of the heavy and light waters have reached equilibrium. The temperature gradient in the cross-section itself, which in cross-section (*2-2¹*) was previously negative from the left bank to the right, is reversed and becomes negative from right to left - for with the constant increase in the inward curvature of the current-axis towards the right (fig. 15) a flow of lighter and slower water of a higher temperature is created to the left. At the point in the cross-section where the temperature gradient reverses, a ford (cross-section *3-3¹*) is established through the weakening of the tractive force (due to transfer of energy from the heavy water to the light water on the right bank). If the profile of the river is compared with the respective energy line, it can be seen that both contours are similar.

The formation of river bends occurs mainly where greater fluctuations in temperature, enhanced by climatic conditions, occur within short periods of time - as in the case of the debouchment of a river from mountains onto the plains. On the other hand, a straight stretch of river with regular, bilateral deposition of sediment is formed where the temperature gradient remains positive over long stretches of river for the greater part of the year.

Fig. 2 shows a stretch of the River Tepl shortly before it flows into the Eger. In this stretch the temperature gradient is always positive, because the water, previously heated upstream by the inflow of hot springs, cools off en route. Over this stretch the Tepl in every respect exhibits a straight course with regular bank-formation on both sides.

The Influence of the Geographical Situation and the Rotation of the Earth

Apart from the influence of terrain and temperature gradient outlined above, the geographical location and the rotation of the Earth (geostrophic effect) also decisively affect the development of a waterway.

By and large, the influences arising from geographical location are expressed in the development of the temperature gradient. In Sweden [far north], for example, the regular climate favours a positive or only weakly-developed negative temperature gradient. The flow of water in rivers is uniform, as is the transport of sediment. The riverbed is perfectly regular, and in most cases trough-shaped (see fig. 17). Heavy water-masses only adjust slowly to climatic conditions of valley floors, and water temperatures are preserved for a long period. Such conditions are also to be found in other mountain streams flowing through cool ravines or forests. Despite enormous fluctuations in the amount of water discharged and the generally steep gradients, moss attaches itself to the stones in such streams. The moment such a channel is exposed to direct light, the covering layer of moss disappears from the stones, which subsequently will be dislodged, and breaches in the riverbank will occur: the channel immediately assumes the character of channels whose temperatures fluctuate continuously.

The earlier a watercourse is exposed to direct sunlight (through clear-felling and clearance), the faster the time and the shorter the distance in which the equalisation of temperature occurs. As a result, the water-masses decelerate abruptly in sharp brake-curves, and transported sediment is deposited prematurely (loss of energy and velocity with the rapid transition from a positive to a negative temperature gradient). Very wide channel-beds are formed so that the water flowing through them under normal conditions is increasingly exposed to the effects of higher temperatures. The immediate result is excessive evaporation and over-saturation of the atmos-

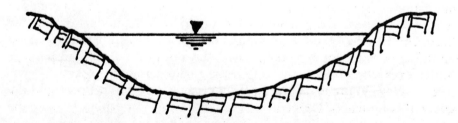

Fig. 17: Symmetrical trough-shaped channel profile

phere with water vapour, which promotes protracted rainfall or sudden catastrophic downpours with the onset of low temperatures.

Venetian rivers enter the upper Italian plain from a steep and almost sheer range of high mountains. Because of this they are subjected to extraordinarily large and abrupt fluctuations in the temperatures of their immediate environment for the greater part of the year. As long as the river continues to flow in the mountains, the water and its surroundings are maintained at a uniformly low temperature. Fluctuations occur only within narrow limits. The morphology of the riverbed exhibits no particular deviations. This all changes the moment the river enters the plain, which for the greater part of the year is warm, periodically hot, and is prone to sudden, strong fluctuations in temperature. Daytime and night-time temperatures also vary by up to 10°C (18°F). The profile of the stream-flow takes on a very characteristic form; a very flat bed with deeply incised gutters (or even two or more gutters in very wide beds) - a pronounced double-profile (see fig. 18).

As a rule the gutters in the *torrente* are very deep. However, since the stream-bed gradient is slight, the velocity of the water in the gutter keeps within normal bounds. Since forestry in the Italian Alps is in a very poor state - whole areas are barren due to neglect over hundreds of years - when the snow melts, great quantities of cold water reach the hot plain without a transitional phase. The ensuing almost instantaneous reversal of the temperature gradient provokes the deposition of large banks of boulder-gravel, which is ejected mechanically by the massive volume of water in the stream-bed - and where the channel is insufficiently wide, this causes considerable flooding.

Rivers in western parts of the upper Italian plain have a completely different appearance, although topographical conditions are the same as the

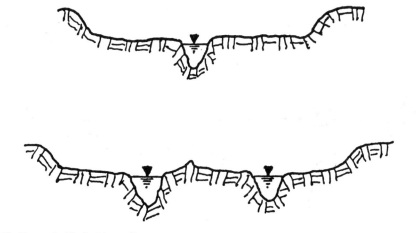

Fig. 18: Channel with double profile

Venetian. The rivers exhibit no *torrente* character, but flow in a regular profile at a uniform velocity towards the river Po. This regularity is caused by the large reservoirs of the upper Italian lakes, which detain the snow meltwater and release it at a temperature already more suited to the plain, so that the formation of such extreme negative temperature gradients, which occur with the *torrente*, can no longer happen.

In northward-flowing alpine streams, conditions are similar to those described above, but not as pronounced as those of the *torrente*, because the northern slope of the Alps is gentler and fluctuations in temperature are smaller. Here, after leaving the mountains, the streams exhibit an asymmetrical deepening of the channel with a build-up of shallow gravel beds on the inside curve (likewise a double-profile) - also a result of the negative temperature gradient present in the longitudinal and transverse sections for the greater part of the year (fig. 19).

In the above, two extreme cases (Sweden and Italy) were discussed. Between them there is of course a wide range of intermediate stages which would take too long to elaborate here. It should be mentioned, however, that rivers which flow into the sea under a positive temperature gradient (those flowing into the Arctic Ocean) carry their sediment far out into the sea (promontory or *haff* formation), whereas rivers discharging into the sea under a negative temperature gradient deposit their sediment prior to reaching it (formation of deltas).

In the case of a west->east direction of flow, the former rivers migrate laterally northwards due the constant enlargement of the heavy water side and the migration of the stream-flow axis towards the northern bank. In the latter case the rivers are widened perpendicularly to the direction of flow in proportion to the decrease in tractive force.

Through the formation of the previously-described heavy water and light water sides, and as a result of helical inversion of the respective water-strata (see *Temperature Gradient, Riverbed Slope and River Bend Formation* on formation of fords), centrifugal effects are induced. These are either strengthened

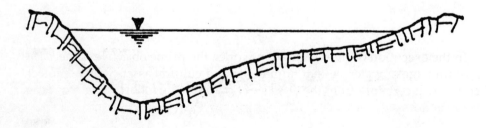

Fig. 19: Channel with an asymmetrical profile

or weakened by the Earth's rotation (geostrophic effect) according to the direction (orientation) in which the discharge of water occurs. Channels flowing in an east->west direction have a different character to those whose flow is west->east, north->south or south->north. In a west->east channel the transport of sediment will be distributed evenly over the whole cross-section, whereas in south->north and north->south channels the transport of sediment is mostly one-sided. West->east and east->west channels will generally be fertile on both banks (although in the latter case both banks will eventually become barren). South->north and north->south channels in the main are fertile on one side only and typically exhibit an asymmetrical deepening of the channel bed.

The General Tasks of River Regulation

In connection with the previous explanations, the following factors are decisive in the formation of the channel cross-section, the development of the longitudinal profile and the horizontal course of a river:

- the topography;
- the temperature gradient ;
- the geographical location;
- the rotation of the Earth.

The topography is dictated by Nature. Where it is essential to protect objects of cultural value, it is possible to use minor retaining walls, although it would be wrong to attempt to regulate a river by means of its banks - in other words, merely to combat the *effects*, but not the causes themselves. In particular, bank-rectification in the form of straight, smooth walls is often dangerous, since the ensuing increase in velocity along the smooth walls will produce the circular motion described in *Temperature Gradient, Riverbed Slope and River Bend Formation,* fig. 16, promoting breaches in the riverbank in a downstream location. A more promising direction for river engineering is *a priori* to regulate the temperature gradient, for with the regulation of the temperature gradient with only minor subsequent assistance from the riverbank itself, the geographical constraints can to some extent be catered for.

In the execution of river regulation works, the prime objective is the *harmless* drainage of water, so that human life and cultural assets will be protected with all certainty from the effects of flooding. The following factors must be taken into account in all river engineering:

a): the longitudinal profile and the horizontal course must be brought into harmony;

b): the channel profile must be so constituted as to enable the faultless discharge of a certain maximum quantity of water in a manner suited to local conditions;

c): precautions must be taken to ensure that water from catastrophic rainfall in the catchment area does not immediately become surface run-off;

d): endeavours must be made to regulate the transport of sediment in such a manner that deposition or removal only happens where desired.

In connection with a); Over the course of time a bed-gradient will be established in a river, related not only to the mean annual discharge, but also to the temperature gradient corresponding to the mean annual temperature. This mean streambed-gradient can then be maintained or engineered through the regulation of the temperature gradient appropriate to prevailing climatic (temperature) conditions. Furthermore, when modifying the longitudinal profile to suit the actual situation, care must be taken to ensure that the sequence of river bends is correct and that, for example, a left-hand bend does not occur where Nature demands a right-hand one.

Referring to b); the channel profile must be adapted to the local conditions and must be capable of an orderly discharge during periods of low and high water flow. The phrase 'suited to local conditions' will be used to mean: in those stretches of rivers which exhibit, and whose nature is favourable to, a natural positive temperature gradient for the greater part of the year, a simple trough-shaped profile would be appropriate. However, where strong fluctuations in temperature occur a profile should be selected which, due to its shape, contributes to the longest possible maintenance of low temperatures in the flowing water. A profile possessing these characteristics is the type of double-profile which rigorously follows the prevailing conditions. In this a natural separation between heavy and light water occurs - therefore drainage of water will be orderly and lateral oscillations of the central axis of the current will be reduced to a minimum, since this will be displaced from the surface down to the deeper part of the channel.

Through the distribution of weight *vertically* instead of laterally the flow of water at the bends is consistent with that of a healthy channel. It prevents a change in temperature gradient within the cross-section, as was described in *Temperature Gradient, Riverbed Slope and River Bend Formation*. Heavy water flows in the lower part of the profile, local conditions permitting, and light water in the upper part. At the interface between the fast-flowing heavy water and the slower-flowing light water, a train of vortices with horizontally-disposed axes is formed, which acts counter to the direction of the current (fig. 20a). This train of vortices distributes the suspended sediment evenly to the right and left of the heavy water core (fig. 20b).

The light water flowing above the heavy water protects it from excessive

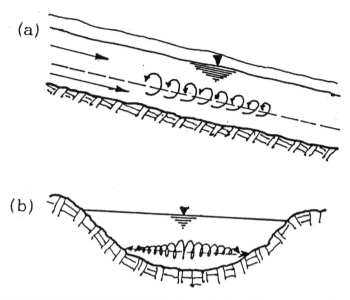

Figs. 20a & 20b: Longitudinal and cross-sections of horizontal transversely disposed vortices in a channel

direct heat. Through this the temperature gradient is maintained for as long as possible in the flowing water. The advancing cold-water core is braked mechanically due to the increasing velocity of the heavy water core, the vortex-train is enlarged and the cold water core diminished, automatically reducing its translatory energy. Conversely, with a decrease in the riverbed gradient the translatory velocity slackens, causing a reduction in the magnitude of the vortex-train and its braking effect.

The correct positioning of this vortex-train is extremely important. The mechanical formation of the transverse profile is dependent on it. In healthy waterways, apart from slight variations in river bends, the axis of the vortex-train lies horizontally, while under abnormal conditions it is sharply inclined or even vertical, giving rise to irregularly-shaped profiles. The appearance and power of such vortices at the interfaces between different velocities are described by Forchheimer:[51]

Where a shallow strip borders on a deep bed, as is often the case where the flow overtops the riverbank, the unequal velocities create vortices with vertical axes. These vortices can excavate longitudinal gutters in the upper riverbed close to the edge of the deeper bed, which have the appearance of pipeline trenches.

During the flood of 14th July 1913, a longish gutter 0.3m-1.5m (1-5ft) wide and 0.2m-1.5m (7in-5ft) deep was formed in this way in the Leonardbach at Graz, by a vertical vortex about 30cm (12in) away from the edge of the deep bed (see fig. 21).

[51] *Hydraulik*, 1914, page 499.

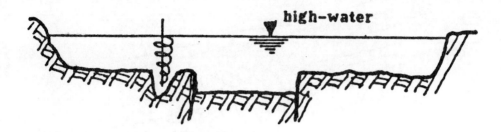

Fig. 21: The formation of gutters in channels

If it is impossible to implement a double-profile (because of too high a cost, for example), then by means of a properly-operated reservoir the discharge in the channel can be structured automatically in such a way that the temperature gradient becomes positive or only weakly negative along the stretch of river in question. In this case heavy water always moves down the centre of the river and the even deposition of sediment and suspended solids on both sides acts to build up the riverbanks, as was mentioned earlier in the case of the Tepl. In this instance the water carves out the appropriate profile unaided, and in the course of time a correctly-positioned double-profile (endowed with the previously described characteristics favourable to the discharge of water) will come into being automatically, a process which naturally takes quite a long time.

In connection with c); the essential measures for preventing rapid stormwater run-off have already been addressed in *Temperature and the Movement of Water*, so that any further comment here would be superfluous.

Regarding d); the tractive force, the sediment transport of a river and their relation to the temperature gradient have already been covered during this discussion. Through the orderly introduction of the colder energy-water[52] present in affluent streams the positive temperature gradient and thus the tractive force can be maintained in the main channel - an objective which can also be achieved through the after-release of low temperature bed-water from the dam. The intensity of the effect, however, will depend on the ratio of the after-flow water to the tractive-force-deficient main-channel water.

The temperature gradient at the confluence of a secondary stream with a main stream must be properly established, otherwise unwelcome phenomena may occur in the main stream in the same way that incorrect regulation of the secondary flow can also play the most appalling havoc in the main channel.

---

[52] The German expression here is *Energiewasser*, which could be interpreted as energised water or water with higher dynamic energy, kinetic energy, vitality, energy-content or a combination of all or any of these. In this regard Viktor Schauberger has also related elsewhere, that when 1 litre (1 Kg) of good springwater at +4°C is drunk, it adds only 700-800 grams to the overall weight of the drinker, the remainder being transformed directly into energy. - Ed.

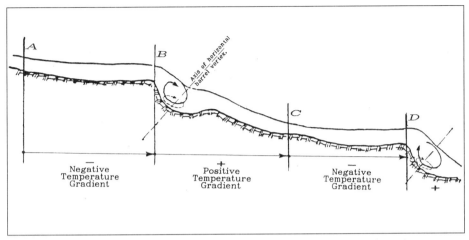

Fig. 22: The alternation between positive & negative temperature gradients in a channel

In this connection attention should also be drawn to phenomena related to the tractive force in unhealthy rivers. As was previously seen in the case of the *torrente,* when cold water-masses reach a warm valley floor, then the longitudinal profile shown below comes into being. This is due to the temperature gradient which at this point has become negative. From A to B the temperature gradient is negative with a major accretion of sediment at B, the point where the tractive force is weakest (see fig. 22).

Here the water has attained its highest temperature. Due to the back-up at B caused by the deposition of sediment, an overfall with potholes and a train of horizontally-disposed vortices (barrel vortices) is formed immediately downstream from B, creating areas of low temperature (pockets of heavy water) which are clearly identifiable. When the light water passes over the top of this colder heavy water it will be cooled from below and the temperature gradient will become positive over the short stretch up to C and from here the whole process is repeated.

For a regulation to be carried out successfully, the alternation of temperature gradients must now be extended over a greater distance, resulting in a more regular movement of sediment and the re-formation of the stream-bed into *gentler* wave-forms.

The Regulation of Temperature Gradient

The establishment of the correct temperature gradient is only possible under two conditions:
- regulation of the temperature gradient through the construction of an impounded lake;
- maintenance of the temperature gradient through the correct form of profile.

On the first point: where topographical conditions permit and there are no problems with water rights, it is preferable to construct an impounded lake in the highest part of the river catchment area. With sufficient depth the lakewater becomes stratified according to its specific density, the lower-temperature water below and the higher-temperature water above. At the point of out-take, the dam wall can be so constructed that water of the required temperature can be drawn from the reservoir through the automatic mixing of water of different temperatures taken from various levels. This is made possible by means of a movable sluice gate, activated automatically by a floating *caisson* directly exposed to the Sun's radiation and the external air temperature, and which will thus automatically release a greater or lesser cross-section of the deeper water-strata (see fig. 23).

In this way bottom-water can be mixed with surface water as circumstances demand. To this end, final adjustment of the floating *caisson* will be carried out after examination of climatic and other conditions, so that at all times the water leaves the dam at a temperature approximating the prevailing air temperature.

Taking this factor into account, the temperature gradient in the sector of the channel decisive for regulation of the whole watercourse (usually the upper reaches) will become positive, with only a gradual and unavoidable transition to a negative temperature gradient. The point of transition and the progress of the change-over can thus be effected at the desired location, which will be selected such that mechanical influences will produce no adverse effects. The reversal of the temperature gradient no longer takes place over short distances, but over a desired longer stretch - and deposition of sediment will also no longer be precipitate, but will be distributed evenly along this greater length. Through the evening-out of the temperature gradient achieved in this way, only gentle modifications to the riverbed will occur instead of the haphazard dislocation of the channel geometry described previously, and conditions will be created which very closely approximate the mean annual temperature gradient and discharge.

In connection with the second condition above: where impounded lakes cannot be built for some reason or other, attempts must be made to maintain a low water temperature for as long as possible - decisive for a positive temperature gradient - through the correct choice of channel profile. Such a profile was described under (b) above. The greatest attention must be paid to the horizontal development of the watercourse (sequence of river bends). For this reason the deeper part of the double-profile must be correctly positioned in relation to depth and handing (left or right) at the river bend in order to maintain the central axis of the current and the proper alignment of the axes of the vortex-train. If the lower, decisive portion of the profile is properly established, then it also maintains its form and position in loose gravel, as is demonstrated by the gutters in the *torrente*.

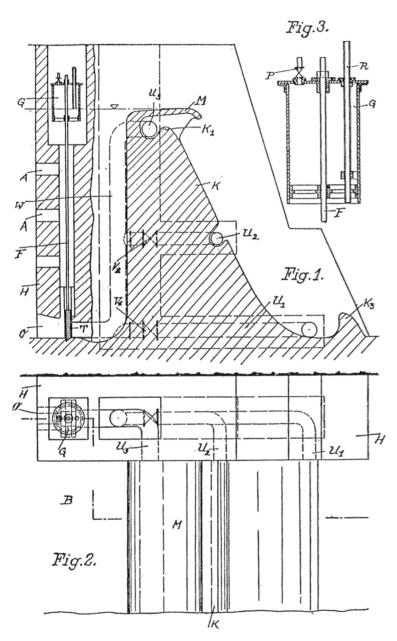

Fig. 23: Patented dam design to control the temperature of the discharge water. (see Patent No. 136214 in appendix)

Conclusion

These are very generalised illustrations of the difficult problems encountered in river engineering and river regulation, when the decisive factors and temperature gradient are taken into account. Detailed explanations can only be applied to specific and individual cases and conditions and cannot be given here.

The perception that mathematical formulae alone are an inadequate basis for the execution of river engineering works, was aptly expressed by the hydrologist Robert Weyrauch - namely, that for the carrying out of river engineering projects,

"An especial gift for hydraulics, an exceptional feel for what is hydraulically possible or impossible is necessary. This is only acquired with difficulty, and even the most experienced repeatedly suffer disappointments."

The Movement of Temperature in Mass-Concrete Dam Walls

[An article by Viktor Schauberger published in *Die Wasserwirtschaft*, the Austrian Journal of Hydrology, Vol. 35, 1930, pp. 717-719.][53]

The fact that increased attention is being given today (1930) to measurement of internal temperatures of large, mass-concrete dam-walls demonstrates the importance attached to these often high internal temperatures and their effect on the strength of the material, on shrinkage cracks and other structural factors.

Temperature surveys at various dams with inbuilt thermometers show that in concrete walls the temperature rises to about 22°C above the ambient temperature within a few weeks, due to the generation of curing heat[54]. In many dams adjustment to the mean ambient temperature takes place more rapidly than in others (Jogne Dam - 1.5 years, Arrow Rock Dam - 5 years). After this has occurred, and with a certain time-lag, internal temperatures

---

[53] The longer passages shown in italics in this section are taken from letters by Viktor Schauberger to the editor of *Die Wasserwirtschaft*, vol.7, 1931, p.7 and vol.3, 1931, p.47 respectively and have been included here for fuller elaboration. - Ed.

[54] *Cast Concrete: Findings in Swiss Dam Construction (Erfahrungen beim Schweizerischen Talsperrenbau)*, Dipl Ing Ed Stadelmann; published by Hoch u Tiefbau AG, Zürich. - VS

[55] There appears to be no actual verb for the process of depositing sediment similar to the process of cavitation (cavitating), therefore the word sedimentating has been coined here. - Ed.

[56] An analogous process is the formation of scale (in boilers), although naturally the *sedimentary* action described above occurs substantially more slowly, since the large differences in temperature in the case of boiler-scale formation are not here present. - VS.

[57] Prof Dr Ing Thürnau: "Die Bewegung der Temperatur in der Sperrmauer der Waldecker Talsperre", *Deutsche Wasserwirtschaft*, vol 4, 1924. - VS

follow external temperatures, but in a more or less attenuated form. This phenomenon will now be examined with regard to its effect on the structure of the dam-wall, with the main emphasis on two aspects:

1): the cavitating, or sedimentating[55] action of water;
2): temperature-induced stresses inside the dam-wall.

Regarding 1): with a simultaneous decrease in temperature towards +4°C during flow (positive temperature gradient), water penetrating into the dam-wall loses its ability to dissolve salts and other substances, or to retain its already-dissolved or otherwise-transported matter in suspension. The tendency to deposit or precipitate these substances will be enhanced by the natural filtering action of the wall-structure, in whose pores the water deposits its dissolved salts as well as some of the accompanying suspended matter (sedimentating action of water).[56]

Water increases its capacity to dissolve matter and to maintain this in solution (the cavitating action of water), if its temperature diverges from +4°C in the direction of flow (negative temperature gradient) as it infiltrates into the dam-wall. Water flowing under a negative temperature gradient and isolated from light and air liberates soluble acids from suffused substances, resulting in a substantial increase in aggressivity and hence in its cavitating action. Before going further, the course of such internal temperatures should be studied using data from the Waldecker Dam[57] in which mea-

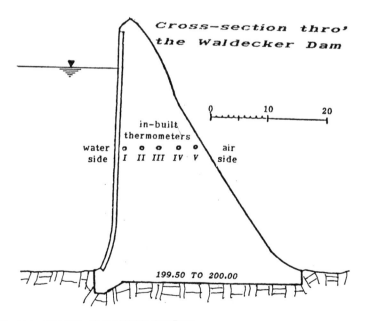

Fig. 24: Cross-section through the Waldecker Dam

surements of temperature were carried out systematically. In this dam remote-controlled thermometers were installed in two different cross-sections about half way up the wall (T_I -T_V - see fig. 24).

The thickness of the wall at the height of the thermometers is 15.3m (50ft). The results of measurement published by Prof Thürnau extend over the period 1914-1918. These begin at the point where temperatures arising from the curing process of the concrete have subsided, so that changes in external temperatures (air and water temperatures) were reflected inside the wall without distortion.

In fig. 25b the course of the external (air) temperatures for the year 1917 are reproduced. In fig. 25a the progress of the internal wall temperatures determined by the thermometers *I, II, III & V* are shown. The temperature graph for thermometer *IV* was omitted, since its curve followed that of thermometer *III* almost exactly. The course of the temperature curves furnishes clear evidence of the relation between external and internal temperatures,

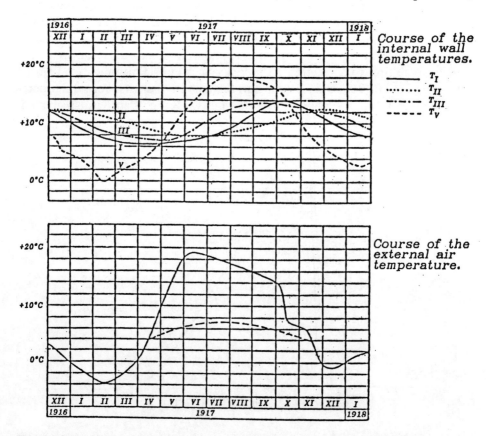

Fig. 25a) Graph of internal wall temperatures   b) Graph of external air temperatures

wherein the wave-forms follow each other chronologically. T_V (temperature graph of thermometer V) follows the curve of the external temperature with only a slight time-lag. T_I, T_{III} and T_{IV} follow the profile of T_V with a lag of about two months, whereas T_{II} exhibits a phase-shift of about 4 months in relation to T_V.

In figs. 26a and 26b the temperature distribution for various days of the year are indicated, from which it can be inferred that from 5th October 1916 the temperature gradient from W (water-side) to T_{II} (water-side to thermometer II) is positive. In the middle of December it levels out to zero and remains negative from then until about mid-July 1917 (greatest amplitude about 2nd May 1917), and then from here until about the middle of November 1917 it becomes positive again. Over the section between T_{II} and A (air-side) the reversed process occurs.

From 5th October 1916 through to 2nd May 1917, the temperature gradient over this section is positive, and from here until the middle of October 1917 it becomes negative. It can be seen that the temperature gradient inside the wall fluctuates markedly, and that in one part of the year (under conditions of a negative temperature gradient) a *cavitating* action of the water will set in, and in the other part of the year (under a positive temperature gradient) a *sedimentating* action will be established. It therefore depends entirely on the conditions of locality, climate and the orientation of the dam-wall, whether the former or the latter action of the water prevails. For example, in north-south oriented walls or in walls exposed to the air on the southern

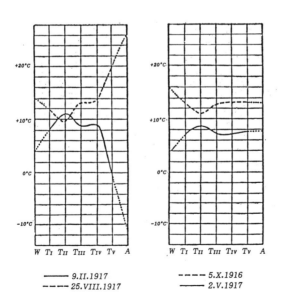

side (east-west wall direction), the cavitating activity (negative temperature gradient) will predominate, as a result of more intensive or more sustained solar irradiation of the air-side of the wall, whereas when the air-side faces north (east-west wall direction) an equalisation of temperatures is also possible.

A wall in which cavitation predominates will exhibit premature 'signs of old age', which manifest themselves as cracks, increased seepage, and so on. In any event the life-span of such a wall will be significantly shorter than a wall in which progressive consolidation is induced through the predominance of sedimentation. This latter case can also be engineered under the most unfavourable local, climatic and site conditions if water is trickled over the full outer face of the dam-wall during the period in question. In this case it is important that the colder bottom-water and not the surface water of the dam is used for the purposes of over-trickling (cooling by trickling of water), roughly according to the very schematic arrangement in fig. 27. With regard to this new type of dam-wall (over-trickling of mass-concrete dam-walls), without going into more detailed explanations, even a layperson can at least understand the purpose and proper arrangement of this over-trickling process, which is required for a few months only.

The over-trickling has nothing whatever to do with the concept of the internal or external stability of the wall as normally construed today. Dams built according to this system will only be placed under water (as are all other walls) when the concrete has cured sufficiently and the wall is stable. The aim of over-trickling is to place a film of water between the external temperature and the air-side of the wall, in the process of which, due to evaporation of water, the wall on the air-side will exhibit lower temperatures than if the Sun were to shine directly onto its unprotected surface.

Through the out-take of cold bottom-water by means of suitably arranged sluices and conduits, water-masses remaining in the storage basin will auto-

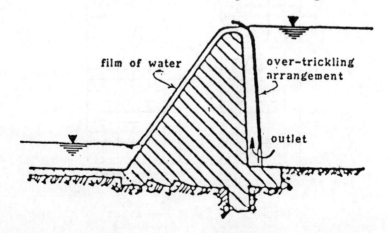

matically be maintained at a higher temperature than the quantities of bottom-water channelled over the external face of the wall. The residual warmer dam-water, in accordance with natural principles, can now readily infiltrate into the unrendered wall. The outcome of this procedure is the creation of a positive temperature gradient in the wall - the water passing through the wall approaches +4°C *en route* from the water-side towards the air-side. As this water infiltrates and becomes increasingly colder with further penetration, the molecules of the wall contract. Becoming enlarged in this process, the wall-pores (voids between the wall-molecules) will be traversed by the water in a positive direction (positive temperature gradient) and according to natural law it precipitates its dissolved matter. Starting at the air-side, because of the lower temperatures on the external face, with progressive deposition towards the water-side the water blocks up the pores (sedimentation).

If at the same time the wall-molecules are reduced to their smallest volume, artificially-enlarged voids are silted up through deposition. If as a result no pores exist in the wall (for all practical purposes), then the presence of any water is also impossible. Signs of fatigue, stresses and temperatures which are mutually intensified by pressure and tension are now also no longer possible. The dam-wall is now immune to the effects of temperature. We are here concerned with similar conditions that can still be found today in the structures of the ancient Egyptians - poreless stones that have remained unchanged for thousands of years, for there too all water and therefore every possibility for movement in the wall have been eliminated.

In fig. 25b the broken line indicates that over-trickling ought to have taken place from about mid-April to mid-November 1917. In this way the air-side of the wall can be insulated from temperatures above about 6°-7°C, and only minor fluctuations in the internal wall-temperatures can be achieved throughout the whole year, which will approximately follow those in fig. 26b. Over the greater part of the year the temperature gradient will be positive and during the remainder only very weakly negative. In this case greater fluctuations in temperature gradient, roughly akin to fig. 26a, are completely eliminated. Therefore the condition where the water's sedimentating action predominates can always be engineered artificially - and through this the progressive consolidation of the wall.

The constant maintenance of a cool exterior wall-surface through over-trickling offers even further advantages: residual curing temperatures of the concrete can be eliminated in a relatively short time. Variations in temperature at the outer, air-side of the wall - often in excess of 40°C (104°F) during the year - will be prevented, hence averting the formation of surface hair-cracks, which foster the destructive action of frost.

Regarding (2): As is commonly known, the action of frost on rocks and wall-structures is based on the fact that the volume of water is least at +4°C. With the onset of frost, the formation of ice (0°C) *hence* triggers off strong tensile stresses in surrounding rock, owing to the increase in volume associated with 0°C water, which can lead to the fragmentation of the rock itself.

Changes in the volume of water under temperatures other than 0°C also play a role, however, when it is remembered that water's spacial coefficient of expansion is about 4.5 times greater than concrete. The effects of these changes in volume are not as obvious as those of ice, and naturally take effect more weakly and over longer periods of time. In the following, a few observations on this aspect are worthy of note.

If a body can expand freely, then its particles of mass as well as the pores between them increase in size. If the body's free expansion is impeded, then in conjunction with the increase in volume of individual particles of mass, a decrease in volume of pores also takes place. This latter case exists in dam-walls, since the particles of mass in the core of the wall will be prevented from expanding freely due to the weight of the overlying sections of the wall (see fig. 28).

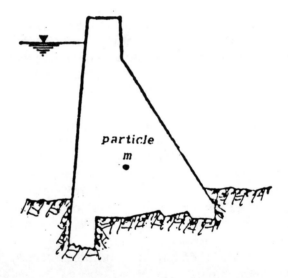

In fig. 29 a series of mass-particles inside the wall are shown. For the time being it is to be assumed that the temperature of the wall increases in the direction of flow (from the water-side to the air-side) - $t_1<t_2<t_3$ and $t_4<t_5<t_6$ (negative temperature gradient). The water-particles flowing across horizon *a-a* therefore experience a rise in temperature from t_1 to t_3 over the distance d, wherein temperatures t_1 to t_3 are created through the reciprocal action between wall and water temperatures. An analogous process occurs at horizon *b-b* where generally speaking the existing temperatures t_4 to t_6 will dif-

The Movement of Temperature in Mass-Concrete Dam Walls

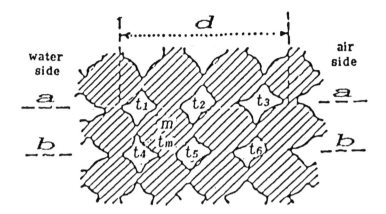

fer from those of horizon a-a, albeit minutely. This is because the processes of friction between the water and the wall in horizon *b-b* during flow will generally speaking be different from those in horizon *a-a*, due to the inhomogeneity of the wall structure.

The wall-particle *m* (see fig. 28), for example, is therefore affected by the various temperatures (t^1, t^2, t^4, t^5) and its self-temperature t^m will experience a small change $\triangle t^m$. However, if $\triangle t^m$ is positive, then the particle will expand and, due to its position in the wall, will do so at the expense of the spacial volume of the pores. The water present in the surrounding pores cannot flow away as fast, as this expansive movement occurs, and the water comes under pressure (the later flow of water from the water-side to the air-side becoming restricted owing to the increase in the volume of water-particles associated with the rise in temperature)[58].

If the same process occurs with a greater number of neighbouring wall-particles simultaneously, then to a certain extent the sum total of these events can eventually lead to tensile stresses in the wall. The constant repetition and intensification of this process during the period of a negative temperature gradient can give rise to unwelcome stresses and signs of fatigue in the wall-structure (for example, in the above example of the Waldecker Dam in the section T_{II} to *A*: May-October).

If on the other hand temperature decreases in the direction of flow ($t^1>t^2>t^3$ or $t^4>t^5>t^6$ - positive temperature gradient), then $\triangle t^m$ will generally speaking be negative. The volume of wall-particle *m*, for example, will be reduced, and since the previously-described increase in volume occurred at

---

[58]Even in cases where the water-body is able to expand freely, with a rise in temperature the pressure of the pore-water against the pore-walls is inevitable, since, as mentioned earlier, the coefficient of expansion of water is 4.5 times greater than concrete. Pressures must become all the more intense when in the examined case the *pore volumina* do not increase in size with an increase in temperature, but actually become smaller. - VS

the expense of the *pore-volumina,* the latter will become larger again. Furthermore, because they are flowing under a positive temperature gradient - their temperature is reducing - water-particles themselves assume a smaller volume. Thus with a positive temperature gradient the conditions are eliminated that could lead to stress-differentials in the wall-structure.

Once again, the way to reduce the previously-described harmful, internal temperature-stresses to a minimum is to over-trickle the air-side of the wall in question, through which high external temperatures can be prevented from reaching the interior of the wall-structure. In this way large variations in temperature within a given cross-section (see fig. 26a for example) can virtually be reduced to zero. As a result, the phenomenon clearly apparent in fig. 25a will also be avoided, since during the period of maximum (or minimum) external temperature, the internal temperature of the wall will nearly be at its minimum (or maximum).

The extraordinarily favourable effect of over-trickling the air-side of the wall during the hotter part of the year should well be evident from the above explanations, since it affects not only the consolidation of the wall through sedimentation, but also insulates it from larger variations in temperature. The increased cost of equipping a dam with an over-trickling system is minimal in relation to the overall cost of construction, and is economic to the extent that it can actually offset the costs of waterproofing the dam-wall.[59]

The possibility that over-trickling may entail unwanted losses of water is of minor importance since, firstly, such losses are only minimal, secondly, the efficacy and lifetime of the installation will be substantially increased, and finally, over-trickling is only necessary until such time as complete self-sealing of the wall through sedimentation has occurred. It should be noted that over-trickling is a one-off affair and is required only during the period in which consolidation of the wall proceeds. The duration of over-trickling should at most amount to six months. Premature removal of this curtain of water or its aeration is out of the question, since only very small quantities of water are involved and furthermore, over-trickling naturally will only be carried out when a sufficient surplus of water is available.

The type of out-take (see fig. 27) suggested by the author offers a further advantage: by mixing bottom and surface water, water of the appropriate temperature can be released as required for the long-range further regulation of the downstream flow-regime.[60]

---

[59] Under certain circumstances, these watertight membranes can actually bring about very unwelcome effects, in that they maintain cement-curing temperatures for an undesirably long period in the wall-structure and prevent the wall breathing. - VS

[60] See *"Temperature and the Movement of Water"* and *"Fundamental Principles of River Regulation"*. - Ed.

Expert Opinion of Professor Philipp Forchheimer

PROF. DR. PH. FORCHHEIMER
WIRKL. MITGLIED D. AKADEMIE
DER WISSENSCHAFTEN IN WIEN.

Wien, den ...15....April........1930.
XIX, Peter Jordanstraße 17

Tel. A 15-8-72

The accompanying design for a dam on the Tepl above Karlsbad embodies features which are new compared with conventional methods of construction. Their description forms the basis of the following report, which consequently concerns itself not with details, but exclusively with the new elements of the design. These consist in the erection of a subsidiary internal wall at a slight distance from the dam wall itself and between these two walls, water is allowed to rise from the bottom of the reservoir. Since the surface water of the reservoir is about 0°C in winter, the heaviest water of around +4°C collects at the bottom; the water temperature thus increases downwards, whereas in summer the opposite distribution of water takes place, namely, its temperature increases from the bottom upwards.

Water gains access to the interior of the wall via the hair-cracks present in all walls. The temperatures of the wall, the water and the air are constantly exposed to large and small fluctuations. The temperature of the infiltrated water always seeks to conform to the wall temperature with the result that its volume either increases or decreases.

With an increase in volume, in the main the internal pore-water advances further and diffuses through the wall, owing to the pressure of water from the water-side. With a reduction in volume and assisted by the suction engendered by the decrease in volume, a further penetration of dam water occurs due to the continuing pressure of the impounded water.

The eventual result of this oft-repeated process - the increase or decrease in the volume of the water present in the pores - is the impregnation of the whole wall with water.

In particular the following process occurs: At the beginning of winter (*Figures I & II* - see fig. 30), with the onset of frost the water in conventionally constructed dam walls freezes to the limit of frost penetration, expands and loosens the fabric of the wall. When warmer weather sets in - solar radiation - the ice melts to the depth of heat penetration, the water emerges from the air-side, taking particles of the wall with it. The danger thus arises that with subsequent freezing, further loosening of the structure will take place. The destructive action of frost would then increase constantly, since the pores will become larger and larger, substantially aggravating the explosive effect of frost.

The way this new dam is designed enables water of about +4°C to overflow the wall and also to enter the upper or the lower trough via the appropriate diverter pipe. In this process the outer wall surface will be protected from the pernicious, fluctuating effects of the external temperature.

In summer it is necessary to differentiate between the behaviour during the day and during the night. In summer during the day *(Figures III & IV)* the water in conventionally constructed dams flushes out the cracks and enlarges the pores, which increases the rate of percolation over the course of time and results in a further deterioration in the condition of the wall.

In the proposed method of construction, use is made of the bottom water, which even in summer is only about +4°C initially and of a somewhat higher temperature later on. This water is permitted to rise between the subsidiary and main walls and to flow over the top of the dam and down the external face. In this way the damaging action of water on the dam will be prevented.

In summer at night *(Figures V & VI)*, while the flushing of the pores will actually be diminished in conventional structures, it nevertheless still occurs. In contrast, with the new construction of the proposed design, no dissolution of the wall-particles takes place.

In Case *III*, even with normal methods of construction and the usual outflow of water via the spillway, the discharge into the drainage channel can be influenced beneficially, because the water cools as it flows downstream, i.e. its temperature reduces from about 18°C to 14°C and hence increases in specific weight. After a sudden drop in temperature, which on occasion can happen in summer, this condition is intensified; the water flows away rapidly with no tendency to form bends. Its tractive force increases downstream, carrying the sediment along with it and the channel bed is deepened. If water is released from the bottom sluice-gate, therefore at a deep level, then the difference between the temperatures of the draining water (about 4°C) and the air (about 35°C) becomes very large. The vorticity of the water then becomes considerable and the formation of bends more frequent, resulting in the deposition of sediment on the inside curve and incipient breaches on the outside curve. With a half-full reservoir, these events diminish and the formation of vortices decreases.

With Schauberger's system the water is discharged at a temperature of about +4°C and the aforementioned processes likewise occur. However, with the aid of the diverter pipes it is possible to select the temperature of the discharge water and thereby adjust it to the prevailing air temperatures in such a way that turbulence and the undesirable formation of bends is reduced.

With regard to the confluence of the hot-spring with the Tepl, the following should be noted. Whereas water of 10°C or 20°C has a specific weight of 0.9997 and 0.9982 respectively, the specific weight of 70°C water is 0.9778; the hot-spring water is thus 0.022 or 0.020 lighter than the Tepl water.

Because of this the mean bed-gradient of the Tepl diminishes due to the interaction between the different specific weights of the Tepl-water and the hot-spring water and therefore part of the hot-spring water initially flows upstream instead of downstream. In contrast with conventional theories concerning flow-velocity, here it is reduced very considerably in the process, retarding the overall drainage of the water and resulting in a corresponding increase in height at time of flood.

It should be mentioned in addition that when a substantial influx of water occurs, some of it is to be discharged directly into the Eger through a fairly large diameter pipe, which branches off on the left-hand side of the reservoir at a high level. This pipe will be rifled, since it has been shown that this produces a sharp increase in flow-velocity.

The Tepl flows into the Eger about 1.6 km downstream from the confluence of the hot spring. At present, where the Tepl joins with the Eger, the Tepl water is warmer than the Eger water. As a result the Tepl enters here at a higher level and therefore no longer has any effect on the sediment, which is left lying on the bottom, damming up the Tepl itself. By constructing the dam and the rifled, high-level, overflow pipe according to Schauberger's design, the Tepl water will in the main be heavier than the Eger water and the accretion of sediment with its damming effect on the Tepl will thus be reduced. This will exert a favourable effect on the danger of flooding from which Karlsbad presently suffers.

From what has been stated above, the superiority of the Schauberger design over contemporary designs is apparent, a superiority particularly evident in relation to the extraordinary conditions at Karlsbad. In any case it is more appropriate to construct dams with adjustable conditions of discharge, whose stability increases with time, than to build dams with a fixed system of discharge, whose structural stability constantly deteriorates, and all the more so, for through the possibility of regulating the conditions of discharge, the necessary storage area and the height of the dam can be reduced.

In conclusion it should be mentioned that Mr Schauberger has already built several barrages (14), which have proved themselves. I have visited some of his constructions personally and I can state that Mr Schauberger's innovations have completely fulfilled their intended purpose.

Forchheimer

134 Fundamental Principles of River Regulation

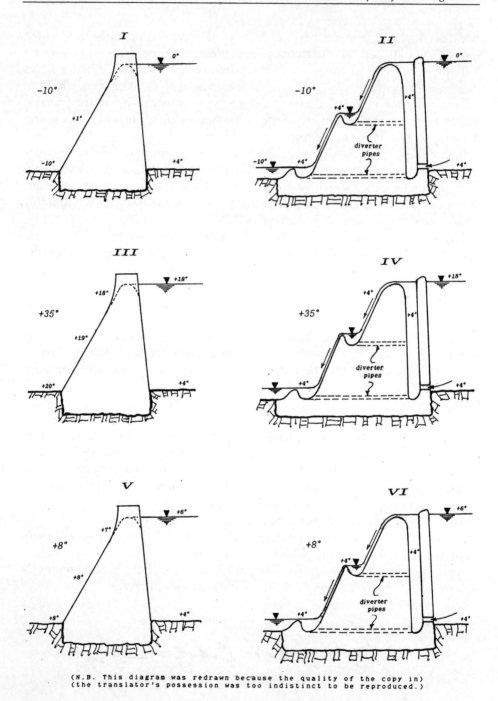

(N.B. This diagram was redrawn because the quality of the copy in)
(the translator's possession was too indistinct to be reproduced.)

Fig. 30: Prof. Philipp Forchheimer's comparison between the designs of conventional dams and Viktor Schauberger's dam that controls the water temperature of the discharge.

The Natural Movement of Water over the Earth's Surface
(The Atmospheric Cycle) and its Relation to River Engineering (Part 1)

[An article by Viktor Schauberger published in *Die Wasserwirtschaft*, the Austrian Journal of Hydrology, Vol. 9. 1931, pp. 131-136.]

Before addressing the actual theme, it is essential to make a few prior comments about hydraulics in general and its implementation in theory and practice.

"The term hydraulics in its broadest sense is understood to apply to all those structures erected in and around water - and in its narrower sense, those structures which serve for water-utilisation or for the prevention of damage by water. River and stream engineering on the other hand encompass all works which enhance the use of flowing water for navigation, and which serve to protect the riverbank against flooding and rupture."

The science of hydraulics further asserts:

"A stretch of river in a state of equilibrium provides the river engineer with the reference point from which to establish a normal profile in a turbulent stretch of river, and to bring about a stable condition. If the flow is confined within the appropriate width (normal width), then it creates the normal profile itself, if it is given the means to achieve this by a skilful hand."

According to these definitions, and in the light of innumerable existing structures, completed experiments and expert opinions, one would be given to believe that the management of water resources is nowadays very close to achieving its highest technical perfection. Every single drop of water would appear to be encompassed by mathematical formulae and therefore in cultivated areas there ought to be no waterway that could deviate by even a centimetre from its prescribed course.

What are the actual facts? What is the practical outcome of an already centuries-old brain-work in the domain of the Science of Hydraulics? Purely and simply, the sorry fact that in all areas of cultivation there is not even a *single* properly-regulated waterway in which a state of equilibrium has been achieved. Firstly, let us take the Danube whose regulation today has already swallowed up almost a million hectares of valuable farmland and enormous sums of money, and will swallow even more - in spite of the fact that navigation is just as fraught as before. To get some idea of the magnitude of this devastation, it should be pointed out that if the Danube *"were given the means by a skilful hand"* to form a normal profile itself, in these areas about 400,000 people could have found a carefree existence. With a deft flourish of the hand, enough land would be reclaimed in order to provide Austria's unemployed with adequate farmland. The same also applies to regulatory works on the Rhine, and the fertile lowlands of Italy and southern France, where already hundreds of thousands of hectares have equally fallen victim to completely misguided river regulation.

Another very instructive example is provided by all our mountain rivers and streams [in Austria], which as waterways are in an exceptionally ruinous state, consuming vast amounts of tax revenue annually. In spite of all this, instead of bringing about an improvement, they exhibit even worse degradation than before, necessitating a constant increase in regulatory works.

The reason why all these present systems of practical river engineering and regulation projects are on, and must pursue, the wrong course is that today **no one knows what water is!**

Thales of Miletus (614 BC) described water, one of Aristotle's four elements, as the only true element, from which all other bodies are created. Not only were the Greeks on the right path towards appreciation of the true significance of water, but they also provided us with information about the practices of long-vanished peoples, whose knowledge must have been considerably more advanced than theirs. Thus in his opus *Timaeus and Critias*, Plato relates that the inhabitants of erstwhile Atlantis regulated their waterways with the aid of *cold* and *warm* water. Only a complete understanding of the nature of water could have brought the Atlanteans to this method of river regulation - a science that sadly was already lost to the Greek sages, despite their intensive preoccupation with the medium of water. Even Thales of Miletus overlooked the fact that the regeneration and further development of bodies with the aid of water is also a question of temperature-associated processes. He failed to perceive the here decisive forces and energies which themselves are a product of the tensions arising from alternating phases of temperature.

In order even to speak authoritatively of the efficient management of water and a systematic build-up of cultivable land, it is first necessary to understand that water and properly-regulated phases of temperature are essential for all new formation, regeneration and further development. Furthermore, if humanity, ignorant of Nature's laws, selects the wrong energy-form for the attainment of its goals, the energy-forms of water will become disorganised and the natural process of autonomous development will cease almost instantaneously and will actually degenerate.

Over the centuries humanity has therefore inherited an incomplete and false conception of the nature of water. Today, in our libraries and archives, we already have a vast amount of literature on water resources management - which bears silent testimony to a cultural advance that regrettably is only illusory. Until the most elementary principles concerning all evolutionary processes of vegetation have been completely understood it is impossible to speak of a real build-up of culture. The development of any culture is directly related to the understanding of its environment - both water and vegetation.

A brief review is necessary in order to understand what is to follow. Every new formation arises from the smallest first beginnings. Further development can only take place if circulation in the interior of the Earth proceeds correctly. In conformity with natural law, higher forms of vegetation build upon the preceding lower vegetation. This new growth is founded on substances contained in earlier vegetation which have been transformed into carbones through the effects of temperature - and which now will once more be decomposed with the aid of water and higher temperatures. In this process of decomposition, water will also be decomposed, resulting in a new mixture of gases which liberate carbon-dioxide[61] as they stream upwards into the suffused and stirred-up salts. With the exclusion of air, this process not only creates entirely new conditions and compounds in the Earth's interior, but also uncovers a new and hitherto-unknown conformity with natural law in the movement of water, which is completely opposite to the presently recognised law governing its movement.

The inner atmosphere of the Earth is created with the aid of water, carbones and temperature - and with further adjustments of temperature it can also dissolve and transport salts. Through the deposition of these salts at the right time and place, the inner atmosphere is able to create a wide variety of new forms of vegetation and new bodies, such as ore and rock - but, naturally, *always under the precondition that the individual phases of temperature take place in the proper sequence.*

From this it is possible to perceive the definite coherence between the vegetation that was formerly present and what is there today. On this basis the interrelationship between all mineral substances is also explained: how the substances are raised from the depths, transformed and refined through thermal processes which take place with the aid of water and its movement inside the Earth, following a hitherto-unknown natural law.

The correct energy-form of water, necessary for growth, was described above, but just what is this energy-form? It is the particular form of water-

---

[61] Since we are here concerned with subterranean chemistry, which takes place under the absence of light, air and sometimes heat, what is liberated in this case may actually be carbonic acid, because the German word used here, *Kohlensäure*, makes no distinction between carbon dioxide and carbonic acid. - Ed.

movement which, through the right mixture of *Sun, Earth* and *water*, results in a sequence of functions that lead to the refinement of primary forms after proper decomposition of basic substances has taken place. From these new and higher forms of vegetation are built-up by the shortest and *straightest* route. In order to explain the functions of the movement of water, it is first necessary to examine the concept of motion itself.

It is generally known that the motion of a pendulum consists of a constant alternation between energy-forms (kinetic and potential). The same phenomenon we also find in the case of electric oscillations which can only arise when two energy-forms, *electrostatic* (capacitative) and *electromagnetic* (inductive), interact. In the movement of water, one differentiates between *laminar* and *turbulent* forms of motion. Laminar motion is the stratified and unimpeded flow of water down an inclined plane. As long as the influence of temperature on the form of water movement is entirely excluded, one can readily speak of laminar motion. However, as soon as temperature is taken into account, any laminar (stratified, ideal) motion is absolutely unthinkable. One can only imagine what such a movement of water would imply: it would mean nothing less than the accelerating descent of water according to the law of gravity, which ultimately (at the lowest point on its path) would have to transfer to a motionless, almost rigid state of rest.

This example shows us how extremes come into being, since a state of absolute rest would then occur as a direct result of water's constantly increasing velocity down the sloping surface - providing a clue to the researcher that he or she is here concerned with a strict conformity with natural law and an orderly sequence of functional processes. Hence the steadiness in the flow of water-masses down an inclined plane (gradient) is solely attributable to the influence of temperature. It thus follows that there can be no stratified, laminar movement of water unless systems of water conduction are specifically directed to this end. This movement corresponds to the kinetic aspect of the motion of a pendulum. It is therefore now quite superfluous to ask whether a second energy-form of water movement exists, corresponding to the potential component of pendular motion.

Turbulent motion then, is viewed as the second form of water movement. Since temperature has so far been excluded as a principal factor, for the same reason it naturally could not be acknowledged as a contributing factor in the proper understanding of the causes of the so-called turbulence of water. To date turbulence has been seen as a vortical phenomenon, attributable to mechanical effects alone - through which various quantities of water of different temperatures are mixed mechanically. A more precise analysis reveals that turbulent phenomena in water are nothing less than the counter-motion to laminar flow - arising from physical causes and generating vortical currents in flowing water, maintaining the steadiness of the descending flow through the creation of transverse currents. Inasmuch as laminar motion is the extreme condition of a real form of motion, the same can also be deemed to hold true for the turbulent motion of water.

In reality we are concerned with two new forms of motion which lie between both extremes and are reciprocally related. In both cases each seeks to approach its extreme condition, but cannot reach it without the intervention of favourable or unfavourable outside influences. For this reason excessively strong turbulence in water must lead to chaotic conditions which express themselves in larger and larger cyclonic storms, catastrophic flood-rains and ultimately in continuous downpours over the same area, while conditions of absolute drought will occur in other parts of the world. It thus becomes clear that in practice we are concerned neither with laminar nor with turbulent energy-forms, but with two other energy-forms: the *positive* and the *negative* energy-forms of water. A positive energy-form (positive temperature gradient) is the internal movement of water temperature that occurs when temperatures of various water-strata approach +4°C, and is therefore a laminar form of motion.

Conversely, the negative energy-form (negative temperature gradient) is the internal movement of temperature when the temperature of flowing water diverges from +4°C (39.2°F). Since a departure from this laminar zero or neutral point occurs when water graduates from +4°C towards 0°C (32°F), then the true zero-point of water is at +4°C, as distinct from all other bodies which contract with cold and expand with an increase in heat. Water expands above and below +4°C - in both instances its volume increases and specific weight decreases. Compared to other bodies, this results in an irregularity, the anomalous expansion of water - hitherto considered of minor consequence, yet playing a far greater role than was ever imagined.

Hence we see that two influences are necessary for either energy-form: the Sun's influence on the Earth and water, and the Earth's influence on water. Both forms of water movement, positive and negative (positive or negative temperature gradients), represent hitherto *unknown magnitudes* in the equation now to be solved - from which are derived not only the great fundamental laws of growth and synthesis, but also the equally important laws of destruction that result in the degeneration of all forms of vegetation. The world is not subject to random accident but is governed according to inner laws, only through the forces of Nature. Left to herself, Nature would have supplanted the earlier vegetation with newer forms, and not only would have transformed the world into a blossoming garden of immense fertility and stable temperature, but in addition would have renewed herself in cycles, as we shall see later.

The opinion that Earth would have been covered with vast forests were it not for humanity's intervention is undoubtedly untenable. Here too, precisely because she would have been left to herself, Nature would not only have withheld the supply of nutrient salts from the vegetation at the appropriate moment, but she would also have ensured that the refluence of sap occurred at the right time. As an example of such self-regulation let us take the beech, when in high summer, due to the development of low temperatures from excessive evaporation in crown foliage, an immediate reflux of sap can be observed. The otherwise unrestricted further development would once again have been regulated automatically by an extremely simple reversal of temperature (change in energy-form).

For this reason optimum conditions, resembling Paradise, must have existed during periods when humanity was still unable to interfere.[62] *Only thus can we explain how extraordinarily fertile soil once existed in a large part of the north coast of Africa, where today wilderness and barren wastes are on the increase. According to the testimony of ancient scribes, in Carthage one could wander all day long in the shade of olive, pomegranate and almond trees. The Carthaginians were delighted to see their vines heavy with grape twice a year and their crops produce more than a 200-fold yield. In contrast to this legendary fertility, reports of the downfall of whole nations through colossal downpours and whirlwinds have also been handed down to us. Paradise and deluge are therefore not to be deprecated as mere fable. These catastrophes and upheavals were initiated by humanity alone, and we are still causing them today.*

The purpose of the movement of water in plants, brought about by reversals in temperature, is to enable the uptake of nutritive material. The clear-felling methods of regeneration initiated by modern forestry must in any case lead to an inevitable and unwanted degeneration, and thus to the initial phases in the death of the high forest. The main reason for catastrophic decline throughout the forestry industry, which more than anything else has led to declining agriculture in upland areas, is none other than an involuntary reversal in the phases of temperature (temperature gradient) caused by the practice of clear-cutting. This results in the cessation of the vital transportation of nutrient salts, which then are inevitably deposited in the wrong places. Light-induced growth[63] is certainly no manifestation of increased growth. It is a fallacy for the detrimental enlargement of the trunk is caused not only by deposition in the wrong places, but also by the deposition of inferior matter, thus preparing the way for future degenerative development. This unnatural growth results in the formation of sinuosities and even in the spiral-like configuration of water-supply vessels - which under normal conditions would lead to the formation of straight, plumb and extremely narrow ducts.

Nature works uncommonly slowly. For this reason it is also impossible to

---

[62] In amplification of the original and similar theme, the preceding italicised text has been abstracted from *TAU*, No. 137, p.22. - Ed.

[63] *Light-induced growth* refers to the rapid and unhealthy increase in the girth of shade-demanding species of timber when overexposed to light, radiation and heat from the Sun. - Ed.

observe the exalted processes taking place in Nature by way of laboratory experiments, since the proper relationships and preconditions are missing. Therefore, even in river engineering, the causes of resulting effects can only be observed in the field in Nature's great examples - in a watercourse from source to mouth. Once it has been determined how a waterway was decades ago and how it is today, records of its continually-changing pattern of flow should first be made, and then *and then only* the causes of its destruction should be sought. Observations over several decades are essential in order to understand the infinitely subtle, constantly-increasing potential in the interplay of forces, which even then is only perceptible through its mechanical effects. The causes, however, remain mostly unnoticed and overlooked, and are usually not taken into account.

This explains why, up to now, we have only ever seen and striven to control the effects themselves. Because of this, all that we have managed to do has been to aggravate the causes - which again leads to the manifold intensification of effects, ultimately provoking catastrophes which legitimately happen according to natural laws. The present critical condition of forestry and agriculture is a typical example of where ignorance and neglect of Nature's laws lead. Without a complete reversal in thinking and approach there can never be any hope of improvement.

The spring that bubbles out of the ground in a healthy forest under the shelter of healthily-grown and undisturbed mother trees has important tasks to perform on its way down into the valley. As it flows down, this water transports nutrients which are destined both for plants, and for the internal constitution and development of animal life. The way in which these nutrients are distributed qualitatively and quantitatively proceeds according to the laws of reaction, without which no life in Nature could exist.

If one now considers the senseless and reckless rape of water in countless wrongly-constructed hydro-electric power-stations, then we cannot even find words to describe the behaviour of the engineer who, in ignorance of the important function of water, thinks only in terms of its exploitability as a cheap source of power and does not concern himself with the extraordinary significance of water in Nature's housekeeping. Moreover, he also fails to realise that with his wrongly designed machines he has destroyed Earth's pulse-beat a thousandfold.

Water's kinetic energy and tractive force is decisively affected by the influence of the external temperature, which also alters water's consistency. The very instant that water comes in contact with the outside temperature, it absorbs oxygen, releases carbonic acid gas (CO_2) and salts are precipitated, which under certain circumstances may be most valuable. Contemporary methods of spring-capture are inappropriate because the most valuable growth-enhancing substances are lost at the spring itself. In metal water-mains the same processes result in an even greater deposition of salts until a more or less stale, insipid and inferior water finally reaches its destination.

The more intense the influence exerted by temperature, the more direct its action and the stronger the effects become. Amongst other important consequences, such as a change in the direction of the energies, this results primarily in the deposition of suspended solids. In addition to the energetic principle operating here (the laws of reaction), mechanical effects will not only be weakened or strengthened by physical causes, but conversely, physical causes will also be strengthened or weakened through mechanical effects.

Of prime importance is the observation of water from source to mouth, taking particular note of the alternating temperature influences and the resulting energy-forms of the water. Water-resources management or river regulation should never be undertaken until it has been established what happens to a drop of water after it infiltrates into the ground and what happens to it in the interim before it rises from the Earth as a spring and flows down into the valley. [The aim of these few lines is to trace the naturally ordained functions that water has to perform along its way. It will thus become apparent on closer inspection that where water comes from and where it goes is of fundamental importance. During the observation of these processes we will be able to establish indisputably that all the changes we encounter in the energy and functions of water are solely to be traced back to the hitherto discounted effects of temperature. For this reason the hasty construction of any large hydro-electric power-stations is to be avoided for there are cheaper methods that produce better results.]

When water evaporates from the sea, leaving all substances behind, the air becomes saturated with water vapour, creating a protective envelope against the direct rays of the Sun. Without this envelope the Earth would inevitably dry up and turn into a desert. Secondly, warming of the Earth's surface is possible only because of the presence of water-vapour in the air.[64] Thirdly, this water-vaporous air is a precondition for a further development of energy (electricity).

It should be noted that formidable climatic changes will occur if, as a result of incorrect systems of forest management and river regulation, the orderly formation of clouds is disturbed. Where these systems have been implemented, the number of thunderstorms has consistently decreased, while those that do occur are becoming more dangerous.

---

[64] The high specific heat of water (lowest at +37.5°C) in normal temperature-ranges for life permits long-term storage of heat. - Ed.

Following from this, the functions of water in the air become clearly apparent. As long as the preoccupation is to conduct water by the fastest and shortest route to the sea, less and less water will infiltrate the ground. This will cause the Earth to cool off, the supply of nutrients to decrease[65] and the development of vegetation and life on Earth to undergo a radical change. If water is to fulfil its task and if the danger of catastrophes is to be averted, it is essential that water is not drained off arbitrarily, but only in a manner suited to its purpose - so that it can carry out its appointed functions according to the laws of Nature, unimpeded. These important processes occur through a truly remarkable interaction between conformities with natural law. All these would be destroyed were water to be channelled according to the dictates of *humanity*, with its supposed laws and mathematical formulae.

The movement of water over an inclined plane (gradient) takes place under a more or less unstable state of equilibrium, which depends on the correct proportion between *water-quantity, gradient* and *temperature*. If consideration is given only to the quantity of water and the gradient, which can be calculated mathematically, the inner energetic processes actually taking place here will never be understood nor influenced in the right way. Neglect of the part played by temperature leads to the destruction of the waterways. It is now very hard to find a place to observe water under the conditions that existed before humanity interfered.

Let us now observe a healthy spring bubbling up under healthy forest conditions. The spring emerges into the outside world with a low water temperature and in deep shade. As it emerges there are found light depositions of matter. In these deposits is a profusion of small creatures which crawl about at the bottom of the spring pool and live off these substances.

Characteristically, water flowing from healthy springs, which are only to be found in healthy high-quality forests, *does not attack the riverbed or the banks, even on the steepest of gradients and despite the often heavy and uneven flow*[66]. In such clear, cold spring waters, bodies (stones *etc*) lying on the bottom are covered with moss and other aquatic plants. Astonishingly, when closely observed, these delicate young shoots hardly move, in spite of the torrential force of water rushing past above them. Careful study reveals that the direction in which these young shoots point changes with a change in

---

[65] Under the exclusion of light and air, deposition of nutrient salts occurs with cooling. Therefore if the Earth cools down, then deposition will occur well below the root-zone of plants, leading to their extinction, and a subsequent loss in transfer of water to the atmosphere, which affects and regulates climatic temperatures. - Ed.

[66] A contributing factor may also be that the water, now mature, has no need of further enrichment by the acquisition of additional minerals and elements. - Ed.

temperature.[67] They point downstream when the external temperature deviates strongly from +4°C, and upstream when it dips sharply towards +4°C. At very particular temperatures the tips of these plants and mosses stand *at right angles* to the direction of the current.

All these observations can only be made in water with healthy temperatures. If such waters are suddenly set out in the open, where clear-felling has been done along the riverbank, this state of well-balanced harmony disappears almost instantaneously. As a result, the watercourse changes colour and character, tears at its bed and banks, loosens sediment, and in areas prone to torrential flows degenerates into a torrent. Whatever measures are undertaken to avert the dangers associated with such channels in the event of elemental disturbances is described as 'torrent confinement'. In reality, however, these merely endow it with even greater dimensions and inaugurate new perils.

The same applies to river and stream regulation. We referred earlier to the regulatory works on the lower Danube, where today 950,000 hectares have been transformed into a flood plain - which have thus become useless as agricultural land. In the *Neue Freie Presse* (New Free Press) Professor Vidrasku wrote: *"Unfortunately, in the belief that construction of huge dams on the banks of the Danube would suffice, works were begun without having studied the problem in sufficient detail"*. He continues: *"If we were to eliminate flooding with massive, high dams, and no longer permit inundation of the flood plain, then the floodwater would rise so high that all our river-ports and all riparian settlements along the Danube would be under water. Moreover, river and lake navigation, which even today is unsatisfactory, would be significantly impaired, and the dams themselves would offer no additional security."*

Another excellent and informative example of intolerable artificially-created conditions is the raising of the Rhine bed at Salez. The cross-section of the valley shows that the Rhine flows along an elevated strip of land, its bed lying up to 4m (13ft) above the lowest level of the valley, and its 1890 floodwater peak 7-8m (23-26ft) above the lowest point of the valley, ultimately intersecting the rooftops of the villages in the lowlands. With good reason Mr Otto Rappolt, a chief government surveyor, states in his book *River Engineering (Flußbau* - Göschen Sammlung) that:

"The deforestation of the catchment area and the system of river realignment is to be considered the principal cause of this hazardous predicament. Thus every important river in Europe provides us with sufficient cause to consider the consequences

---

[67] Many people may perhaps have already noticed that a small boat, anchored to the bank, shifts its position with the Sun, and can even float upstream, broadside to the current, or that a vessel of almost any description (from a rowing boat to a ship) attached to a buoy always points in different directions during the course of the day. These examples naturally only apply in calm weather. - VS

that incorrect river regulation can draw in its wake. The rivers Etsch, Po and Tagliamento in particular will present the Italian Government with many more problems, as well as swallowing up considerable amounts of capital, if there is not a radical departure in present government policy."

Very soon it will become apparent that these colossal structures were not only totally useless, but have also initiated damage on a scale that even the competent authorities find hard to assess. If traditional systems of torrent-confinement, river and stream regulation, and contemporary methods of hydroelectric powerstation construction continue to be used, the causes of the increasingly frequent catastrophes will never be eliminated. On the contrary, these incidents will assume even greater magnitude from year to year. The havoc caused by the destruction of our high forest has without doubt already tipped the scales.

The Natural Movement of Water over the Earth's Surface

(The Atmospheric Cycle) and its Relation to River Engineering (Part 2)

[An article by Viktor Schauberger published in *Die Wasserwirtschaft*, the Austrian Journal of Hydrology, Vol. 10. 1931, pp. 148-153.]

In the sheltered environment of forest a waterway can maintain its steady temperature to a certain extent. The external temperature exerts only just sufficient influence - indirectly, through the triggering of weak vortical phenomena - as is necessary for progressive reduction in tractive force. Healthy forest, or cold affluent streams that join the main body of water along its course, ensure that absorption of heat proceeds slowly, providing a basis for an even release of nutrients for the benefit of flora and fauna in its immediate surroundings. (The Amazon river in Brazil and the rivers of Java, for example.) Without healthy forest there is also no healthy water, no healthy blood and therefore a deterioration in the most fundamental conditions for life, all of which result from the present methods of water and forest-resources management.

The longer water is shielded from the direct light of the Sun (through forest protection, proper arrangement of lakes and cooling affluent streams), the longer it will retain its energy and above all its tractive force, and the more regular the release of its substances, its overall flow-characteristics and direction. Properly managed water - that is, water with a temperature adjusted to prevailing climatic conditions - cannot attack riverbanks, as will be shown later. Incorrectly handled water whose temperatures are generally too high, has the capacity to tear into riverbanks, and for this reason assumes the characteristic behaviour of a torrent. It must be emphasised here that all attempts to maintain the equilibrium of a watercourse by means of the riverbank itself, in the form of bank rectification, are futile. The best proof of this fact are all the regulatory works carried out in accordance with this principle, which, despite the ongoing need for repairs and maintenance, continually give rise to new damage and further expense.

Naturally, the distribution of water-masses according to their various states of aggregation is of special importance. Premature evaporation of water-masses (due to warm soils) during drainage over the ground surface results in disproportionate accumulations of water vapour in the atmosphere. Because of this, a reversal in the reciprocal interaction between temperature gradients occurs too rapidly in the atmosphere, which again must lead to heavy rainfall and cyclonic storms. The ensuing reciprocal effects of ground-temperature produce undesirable temperature gradients in draining water-masses, through which the full hydrological cycle of water is further adversely affected. In such instances, instead of infiltrating the ground, water flows into ever-broader channels. As a result of excessively high temperatures thus arising, water transfers into the atmosphere too rapidly. Because of this, not only is the vital extraction and supply of nutrients arrested, but in this way even greater and more catastrophic floods must also follow.

If we now apply the above concepts of temperature gradient, we are presented with a case where a blocking transverse current flow is created due to an over-strong outside influence, resulting in the widening of the channel through too rapid a departure from +4°C (thermal expansion of water). In such a case we can also speak of an excessive negative temperature gradient, which must lead to a loss of the draining water's state of equilibrium, and to heavy, flow-dislocating accumulations of sediment. More and more the draining water approximates the turbulent second energy-form, entirely loses its inner equilibrium through increasingly stronger counter-movements and eventually begins to scour the channel bed, creating pot-holes and washed-out depressions in order to brake its descent. The channels become wider and flatter, the influence of external temperature more intense, and in the process, larger and larger volumes of water are returned to the atmosphere. The apparent disappearance of the water, the constant increase in dried up channels and in catastrophic storms are the legitimate consequences, rightly attributable to river regulation projects as carried out today.

From the following results of observation it will become clearly apparent that current systems of water management must gradually lead toward total elimination of all species of vegetation, hence all forms of agriculture. The economic decline taking place before our very eyes is but a first step down this road, which only a return to Nature may yet perhaps be able to arrest.

Frightening examples of the result of such treatment of water are deserts, which often were once the abode of higher cultures. Excavations bear witness to the efforts that were made to preserve the dwindling supply of water through large-scale hydraulic installations of all descriptions, which regrettably were poorly designed. Indeed, we have but to glance at a map of the Gobi desert, whose rivers dry up around the edges of this ever-widening wasteland.

Thoughtless treatment of vegetation and excessive influence of the external temperature at these latitudes have created the present pattern of the deserts. Once the reasons for the present destructive phenomena and how they were caused are fully understood, new places to live can be wrested from the deserts by reversing this gradual decline - by rearranging the energy-forms in question - and new possibilities for life can be created by starting again just as gradually from the very first beginnings.

To date the concept of temperature has not been evaluated at all. Water flowing down a gradient is governed by two different influences: the direct influence of solar radiation and the indirect influence of the Earth's moist bulk. Both these influences maintain the unstable state of equilibrium in draining water through changes in the energy-forms, which in turn modify the gradient. Quite obviously the Sun's influence must be greater than the Earth's, and this more potent influence will primarily be brought to bear on the upper margins of the channel body. At this location, due to the influence from outside, water will *exceed its critical velocity* and become more strongly turbulent. If the outside influence is indirect - for example, due to the shelter of the forest - then it exerts an effect on both sides more or less uniformly, as long as both banks are similarly constituted. If the outside influence acts directly (direct solar radiation), then it gives rise to an irregular, non-uniform development (diurnal fluctuations). This greater effect of temperature results in the formation of sharper or gentler bends in the river.

In a manner of speaking, accretions, incipient breaches and other symptoms of flow-dislocation are to be viewed as places selected by prevailing disturbances for the deposition of matter. Once again, particular attention is drawn to the fact that deposition occurs *in the interior of the Earth* (with the exclusion of air) as the temperature *approaches* +4°C, whereas *on the surface of the Earth* (under the influence of air) this deposition takes place as the temperature *moves away from* +4°C.

Apart from this important physical outside influence, consideration must also be given to the influence of frictional heat (mechanical effect). Cold, healthy mountain water flowing rapidly over the riverbed and along the walls and edges of the riverbank generates slight eddies and hence counter-currents at the above contact surfaces. This is as a result of localised exceeding of critical velocity relative to water temperature, induced by frictional heat. The prerequisite for such a counter-current is a difference between the influences of the outside temperature and that of the ground.[68]

The lower the temperature of the central mass of core-water, the greater the relative rate of flow. In the same relative proportion this now triggers off a reaction on the riverbed and banks through mechanical and physical impulses in the form of counter-currents. If the proper proportions between temperature, mass and riverbed gradient should now exist, then the preconditions for the unstable state of equilibrium of the water-masses would be met. This state, however, is virtually impossible. If the riverbed gradient is too slight for the prevailing water temperatures, then removal occurs - or in the opposite case, accretion. The riverbed gradient therefore adjusts itself according to the temperature gradient, whose constancy will again remain undisturbed as long as protecting forest continues to exist in its proper measure and composition (presence of cold affluent streams).

The preservation of the watercourse is thus exclusively dependent on the proper conservation of the forest, and the necessity to regulate waterways today is therefore a consequence of unnatural forest management. Should it now be contended that the forest would in any case have to be cleared for agriculture, then it is to be countered that this is quite in order up to a certain extent, on condition that suitable substitutes (properly constructed reservoirs and river profiles) can be provided in lieu of forest. In the absence of forest these substitutes must be capable of constantly maintaining the essential stable state of river equilibrium.

Under such preconditions, the previously-described phenomena observed in the behaviour of aquatic plants will also be reproduced. If these plants should now incline up- or down-stream, then it is merely the effect of slight fluctuations and rearrangements of the temperature gradient. If the tips of mosses stand motionless and perpendicular to the direction of the current, like a needle on the scales pointing towards zero, then in this way they affirm the existence of the proper conditions for equilibrium. The formation of minor eddies (counter-currents) described above continues to take place in the proper measure for as long as the correct conditions of temperature prevail in the main body of water.

If, after the removal of forest, the water is exposed to the direct light of the

---

[68] It could be construed from this that if water, riverbank and air temperatures are equal, then the tendency for the formation of such counter-currents would diminish. Thus, if the riverbank and air temperatures were to cool in step with the increase in flow-velocity then, natural obstacles apart, turbulence due to the exceeding of critical velocity relative to water-temperature would be reduced to a minimum. - Ed.

Sun, then along the upper contact-surfaces of the riverbank strong vortical currents (turbulence) develop. The central core water-masses forge *ahead and exceed their critical velocity.* The gurgling up of turbulence in these core water-masses at the position of the greatest velocity is an after-effect whose original cause has so far not been clarified. In a manner of speaking it is the emergency brake against over-rapid drainage of water-masses down a gradient, and the hitherto unexplained maintenance of the steadiness in the flow of water down an inclined plane (gradient).

All of a sudden a reaction sets in, in the form of a sharp counter-motion. In this braking curve the whole body of water will be pulled around hard to the left or to the right. Pot-holes in the riverbed develop because the runaway water is braked too severely and too abruptly, subsequently leading to familiar destruction of the riverbed and banks. Instead of being alleviated or arrested, these phenomena will only be made far worse through contemporary preventive measures. It should be mentioned here, for example, that avalanches are also created in a similar fashion, and in the process of combating their ravages, as is the case with water, mechanical effects alone are perceived - and with no consideration given to the physical causes, attempts are made merely to redress the mechanical effects themselves.

A further function of the formation of vortices, which evolve through mechanical and physical processes, is to enable aeration of water-masses, adjustment of water temperature and modification of kinetic energy - as a direct result of which banks and riverbed will also be re-modelled at the same time. *The riverbed gradient is therefore a secondary effect of the temperature gradient.*

The transverse, blocking disposition of the water-masses, the widening of the channel, the external influences which thereby exert a more direct effect on the one hand, and the eventual evaporation of water *en route* (drying up of rivers), the creation of deserts *and* unparalleled floods on the other, are the all-too-familiar outcomes of manipulation of water according to current theories. This end-result, however, is the entirely legitimate and proper consequence of contemporary systems of water resources management. Disasters and devastation must therefore increase automatically in precisely the same ratio that capital is sacrificed for river engineering projects carried out according to conventional practice.

In the above, the influences affecting water have been described in broad outline. In the following it will now be shown in greater detail just how dependent aquatic creatures are on water, and how they have to pay the penalty for every man-made mistake.

In healthy water immediately below a spring, we find the healthy mountain trout, famous for its tastiness. Under the careful scrutiny of a watchful observer, this stationary trout, which enjoys a comfortable and peaceable existence in healthy water conditions, is reminiscent of the gentle swaying movement of the previously-described moss-tips. To those who know how to observe, this now offers a plethora of fascinating insights. They learn to understand the purposefulness of the trout's very slightest movement and begin to realise that both in theory and practice the human mind consistently has the perverse inclination to take the wrong road, although Nature constantly demonstrates the right way with countless reiterations and allusions.

With the exception of the spawning period, feeding is the sole reason for every movement. Slight changes in the height or depth of station neutralise the variations in the movement of the food supply, occasioned by outside influence. If the trout is frightened, then it bolts upstream like a streak of lightning, only to return once more to its former position after a certain period of time.

With a very simple procedure it can be established that, as a rule, a trout positions itself in the axis of core water-strata at the place where water particles closest to +4°C flow, which are therefore least turbulent. Water particles moving along this axis possess the relatively highest velocity, due to orderly (more laminar) forward motion. All foreign bodies heavier than water, including the trout's food, also travel along this energy-line, which is the true axis of a river. It is also here that turbulence and eddies created by the trout's own body are best able to assist its own motion. As long as healthy conditions prevail - as long as the core water-masses remain in proper relation to bed-gradient and river bends - vortical phenomena will continue to appear, which are more intense at the upper margins of the riverbank and which become increasingly weaker the further down they occur. Under such conditions the position of the current axis hardly varies at all.

The minor vortices running counter to the current-flow at the edges carve out small hollows and cavities as the channel necessarily widens and flattens out. In the process they dislodge bits of soil from both banks, and with them food for fish (worms which prefer to live in the cool, damp edges of the riverbank). The constant increase in the strength of outside influence, due to the widening of the channel, ensures the progressive reduction of water's tractive force, and with it the even deposition of salts, which are anyhow of increasingly lower grade.[69] Towards evening the temperature gradient once again approaches a positive energy-form and in the night the transport of sediment takes place. Similarly, higher quality water also arrives in the lower reaches during the night (less deposition - greater sediment transport - deeper bed). At an appropriate temperature the watercourse regulates itself entirely. Noticeable bed-load movement, sediment accretions or breaches of the riverbank are unknown phenomena in healthy waterways.

[69] The higher the water temperature, the more inferior the quality of dissolved salts. - Ed.

In waterways with the correct conditions of water temperature, the channel will not only widen itself in the right proportions but will also deepen itself in the lower reaches. Therefore it automatically develops longitudinal profiles suited to the varying shape of the river cross-section - to the extent necessary for the removal of bed-load, which under orderly forest conditions is in any case minimal.

If flooding occurs in watercourses still undisturbed by human hand, then the influence from below becomes more intense as the volume of water increases, due to external temperatures - which as a rule are low at such times. As it also does at night, the temperature gradient approaches the positive energy-form, and because of lower temperatures, the water-masses flow faster without overflowing the banks or attacking them. The temperature gradient is primary in function and the riverbed-gradient becomes secondary if, as a result of proper rearrangement of temperature, the mass-transport of sediment is regulated in accordance with the corresponding increase in velocity.

Were the channel not to flatten out (as in fords), then a reduction in tractive force would also not occur. The release of nutritive matter would also not take place, owing to low temperatures which remain constant in deep waters.[70] In properly managed water conditions - or more accurately, in untouched waterways protected by natural, undisturbed forest - the correct channel profile will be formed. This will also lead to an increase in the volume of water in the lower reaches, and with such an increase in mass, the corresponding influence will also develop from below, from the Earth, resulting in a weakening of the stronger influence from above (the Sun). Because of this, it will not only be almost impossible for sediment to be left lying on the bottom, but it will also be almost impossible for flooding to occur in the lower reaches.

The greater the volume of water, the greater too the rate of flow under otherwise-identical conditions (e.g. gradient), as a result of the correct rearrangement of temperature. With the establishment of a correct profile, the temperature gradient automatically activates the energies required for the transport of such large volumes of water. The build-up of energy required for sediment transport in the lower reaches takes place through the confluence of lower-temperature affluent streams with the main channel.

With current methods of river regulation, the outside influence inevitably becomes more and more direct and therefore increasingly apparent. Peripheral vortices become stronger and the body of core-water surges

---

[70] The radiation of vortex-accumulated energy and the deposition of nutrients and trace-elements occurs at the ford, where the longitudinal vortex slackens, reverses its direction of rotation and the river widens, flattens and deposits sediment, nutrients, salts and trace-elements. - see figs. 4, 5, 6 & 13. - Ed.

ahead with greater force. The critical velocity of the core water-masses will be exceeded on a larger scale through constant repetition. Owing to the formation of pot-holes and striations the current axis will become increasingly ill-defined, the watercourse ever wider and the deposition of sediment greater. Even the trout will only be able to obtain its food by changing its position more and more frequently, since this no longer travels as centrally as before.

Here too, in the same way as for *humanity*, hard labour for its daily bread begins for the fish. With this further development, an increasingly futile struggle begins for its very existence. Even for the otherwise peaceable trout, this untoward event initiates a battle for the survival of the fittest, affording an existence to several species of predator whose life-expectancy at this stage is also very limited, since the watercourse and the water in it are already diseased and in the process of drying up.

If one observes micro-organisms that evolve in the water just below a spring under the influence of light and the Sun (heat), then it also becomes clear why fish seek out certain spots at the head-waters of a stream at spawning time. It is becoming harder and harder for mother-fish to find a suitable place to lay their eggs, and where young-fry, left to fend for themselves, will have a chance to begin their lives. The decline in the former abundance of fish is mainly attributable to the increasing scarcity of healthy water, which instead of gushing forth from noble, high-grade springs, laden with nutrients and trace-elements, trickles out of the Earth in empty seepage springs. Already unhealthy at the spring itself, such water is unable to generate healthy drainage conditions downstream. The core water-masses no longer surge ahead momentarily, and instead of a convergent (centripetal) movement of water-particles, the whole central body of water becomes turbulent. Its energies are directed towards the banks because of accretions of river-gravel and sediment, resulting in breaches in the riverbank and the formation of islands.

At this point attention should be drawn to a major misconception. This forward surge has so far been viewed as a process of continuous acceleration. In reality, this supposed acceleration is precisely the precondition for its inner braking, through the incidence of increased turbulence where the flow-velocity is greatest. Consequently, all mathematical calculations hitherto applied to this phenomenon must undoubtedly lead to exactly the opposite effect (a reaction effect).

Eventually a bend in a river is not only formed horizontally, but also the water actually begins to curl in upon itself. Over-heated surface-water becomes turbulent and trails behind the deeper, colder water-strata, which continue to surge ahead, altering their consistency and eventually falling back themselves. At certain velocities, by mechanical means, ripples are formed, which break backwards in an upstream direction, throwing the food floating on the surface upwards and towards the source. Strong evaporation suffocates insects dancing above the steaming surface water, which then fall in, providing food for trout and swallows.

As long as healthy water conditions prevail, a trout enjoys great abundance and a wealth of choice. In cold, clear water, the stationary trout easily espies every morsel of food floating towards it, alters course with a flick of its fins - and with an expert eye evades the angler's hook. The small fish are the main ones that bite, because they swim outside the mainstream of the current and hungrily seize upon every scrap of food that deviates from its regular path. If sultry weather conditions occur, and if temperatures rise in a sharp curve (negative temperature gradient), then the whole body of corewater becomes turbulent. Now becoming hungry, the stationary trout nervously darts about, lunging at gnats dancing above the surface and becomes careless. It falls easy prey to the angler, who is well aware that in such weather the large trout bite - but has no idea *why*. The same naturally applies after a fall of rain, when newly-arrived water necessarily mixes with riverwater, which becomes strongly turbulent in the process - chaotic flow conditions arise, which make all orderly supply of food impossible, and the trout becomes ravenous.

The principle governing the movement of fish in water or birds in the air is the same, although there are obvious structural differences attributable to different properties of the respective media. In its own way, each medium can be influenced by way of reaction phenomena in order to produce the desired effect - the most efficient forward motion. Like the fish, the bird has the ability either to overcome resistances peculiar to the medium through physical (not mechanical) processes - and without any major expenditure of energy, or at least minimal expenditure at the right moment. What has hitherto been held to be impossible - a large output with a minimal input - will be restored to the realm of fact, shaking the very foundations of current theory concerning energy and its conservation.

The various resistances and the friction which, following natural law, *increase by the square of the starting velocity,* will be neutralised in the same ratio by the physical factors mentioned above. Consequently, by exploiting this new factor at the right moment, all that will remain to be overcome in practice is the resistance-less medium. Once again, the most detailed analysis of the energy-forms here in question is necessary in order to replicate this ideal motion artificially in either medium to suit our purposes.

While it is quite remarkable that trout can move so rapidly upstream, in view of its relatively modest energies, at first view it is equally baffling that the trout is able to remain motionless without any appreciable effort in fast-flowing waters in which a human being can barely stand up. In clear, cold, healthy mountain water, a stationary trout positions itself in the central axis of the current, where the water-filaments flow that most closely approximate +4°C. The trout influences the velocity of water flowing past it with the smooth, slippery surface of its skin - at first with a purely mechanical energy-form. Displaced by the mass of the trout's body, the water initially comes under pressure and slips past the slimy body with a greater velocity than water particles further removed from it. The immediate effect is that the highest flow-velocity possible under the prevailing water temperature is exceeded, producing an increase in turbulence. In the neutral zone created in this way by the counter-flow of the water, the trout is able to stand still without effort (see fig. 31).

If the water becomes too warm as a result of external influences, then the reaction will also be weaker. In particular motion and counter-motion will be thrown out of balance, and the trout will be pushed gently downstream. Conversely, if the temperature suddenly drops, it will be propelled upstream. The trout also quickly vacates places where changes in the unstable state of equilibrium are most extreme and seeks out a new position suited to the changed conditions, so as to re-establish the lively balance between water, reaction (counter-movement) and body-form. The establishment of the state of equilibrium outlined above evolves purely and simply by mechanical means alone, due to an increase in turbulence.

By breathing through its gills and therefore by means of *purely physical impulses*, a trout is able to neutralise, not only the effects of rapid movement of water, but also smaller irregularities. The effects triggered off are much

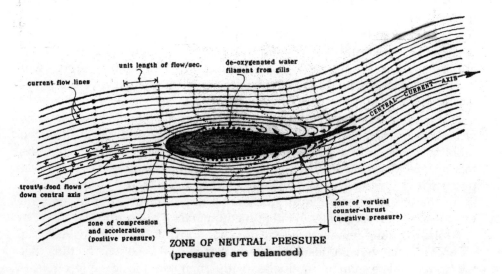

Fig. 31: The Stationary Trout

larger than the trout's relatively minor applications of force, and because of this they can be raised to an unusually high level of efficiency. A trout is forced backwards by flowing water the moment gill-breathing ceases. When the trout swims rapidly upstream, then it breathes with greater intensity and rapidity and, in a manner of speaking, when it takes to its heels, its gills operate at full throttle.

These processes are explained as follows: the formation of counter-vortices described above provide the trout with its raw condition of equilibrium. They evolve mechanically and are suited to the size of the body and form of movement. The instability of the state of equilibrium, however, creates the possibility of increasing the velocity of the water by physical factors, through which increased turbulence is caused by pressure from the gills. Breathing through the gills not only serves for the intake of oxygen, but also for forward motion - an arrangement which illustrates Nature's absolute superiority. The trout increases the velocity of water through deep and rapid breathing, to the extent necessary to conduct a correspondingly-larger volume of water (from which oxygen has been removed) through the gills and alongside the body. In this way, vortices that have been intensified in a potential sense can now be generated in almost any desired form and strength, as counter-movements. In this fashion counter-currents originally evolving through the mechanical formation of vortices will be multiplied and the trout needs only to redeploy its tail-fins in order to exploit the reactive effects. The end-result is acceleration of the body in the opposite direction to the draining water. In the interval between breaths, mechanical vortices alone are active. The gentle twisting movements of the fins destroy the vortices created mechanically (destruction of vortices by formation of vortices) and the trout moves backwards. With the very first expulsion of breath the trout again stands still in the most torrential of flows.

The secret lies solely in the exploitation of hitherto-unknown laws of water movement. Therefore, it will no longer sound strange when the assertion is now made, that *not only are today's hydroelectric turbines both uneconomic and unreliable, but ships and aeroplanes are as well,* since they operate contrary to laws prevailing in their respective medium of air or water. One of the most immediate practical applications, achieved by influencing the medium artificially but according to natural law, will provide proof that it is possible to cause any propeller-driven aeroplane to nose-dive from any desired height. The air-pockets and vertical wind-shear so feared in aviation which force aircraft into a vertical dive, are none other than the interaction between temperature gradients - which as a rule regularly reverse early in the morning or towards evening. Inasmuch as air-crashes can be traced back to these causes, the majority always occur at such times. The time of year also plays a major role. These inversions in the temperature gradients can be engineered artificially by very simple means. Moreover with the aid of this knowledge, one of the most recent and supposedly greatest achievements - air warfare - can be eliminated *easily*.

Present systems of powering aircraft with propellers operate against the laws of Nature, and it could therefore be said that flying today is an exceptionally risky game of chance. Indeed, it could almost be said that even one untoward ray of sunlight could bring it to an abrupt end. The basis of this assertion, and the reason why glider pilots can remain aloft at certain hours only, will be addressed in subsequent chapters.[71] Since aviation is still in its infancy, and in view of the limited experience gained over the few years since its inception, this should not be taken as a reproach.

If, on the other hand, we examine what is perhaps the most ancient of technologies, the rafting of timber, and if we also consider that this method of timber transportation is still practised today in spite of the destruction and incalculable damage it has wrought, then no word is strong enough to criticise this unbelievable thoughtlessness.

Through the removal of forest, the means of transportation - water - is laid bare to the elements. The inevitable outcome of this is the destruction of the channel. With the channel in such a condition, the controller of rafting operations adds to the problem by releasing even more water from holding basins. What is thus achieved is precisely what was not wanted, or should be avoided. Instead of being carried forward, logs will be thrown broadside to the current and stranded. To guard against this, the forester then creates his embankment works,[72] which incidentally were also adopted by river engineers later on. Revetments, stone encasements, guideways, groins and transverse barriers are now supposed to hold the timber away from the bank.

Briefly the effect is as follows: in company with the water, timber flows along the now smooth, steeply raking embankment wall in similar fashion to water sliding along the slippery body of the fish. Significantly larger turbulences are created. The encased-stone wall will quickly be undermined following Nature's hydraulic laws, and collapse after a short time. As long as these wall-surfaces hold out, the turbulent water-masses will be forced into a completely unnatural flow-form. Because of this, the timber surges ahead too rapidly, and all means of controlling the water are lost. Intensified in this process, counter-forces flow around the end of the reconstructed

---

[71]Unfortunately these subsequent chapters never came, because upon Prof. Philipp Forchheimer's death, under whose sponsorship these articles in *Die Wasserwirtschaft* were written, all further papers by Viktor Schauberger were refused publication - Ed.

[72]The root of the German word *Uferverbauung* employed here is *verbauen*, which can also mean 'to bodge up' or 'to jerry-build', and it is a matter of conjecture what emphasis Viktor Schauberger actually intended ! - Ed.

embankment and demolish it. Having passed the revetment, the accumulated energies attack the now-unprotected bank with increased force and destroy it altogether.

This makes further bank stabilisation imperative. Thus an endless string of revetments are built at enormous cost, which not only make all useful floatation of timber impossible, but also are precisely the most dangerous flow-guides in time of flood. Potential forces are generated and accumulated at these guiding structures. Far below in the valley, where no dangers existed previously, destructive effects will suddenly be unleashed which act equally as unexpectedly as they do abnormally, laying valuable farmland to waste,[73] which is scarce enough anyway without this forest-destroying technology.

In fact, because of these measures, a stage has already been reached where, almost without exception, all the once-healthy alpine streams, abounding in fish, are now in a positively ruinous condition and devoid of fish-life. In time of flood these waterway ruins convey vast quantities of sediment down into the valley, causing unparalleled havoc, only to dry up again during normal times. Hydro-electric plants being built today are responsible for all the rest of the damage. If our hydraulic engineers continue with their present methods, and should they eventually seize upon the last high-lying healthy reserves of water, then in a few decades we will live to see the day when all the power-stations, built at vast expense, stand empty. In the process we shall also lose the last vestiges of fertile soil which today, even if already in short supply, is nevertheless *still* available to us.

As long as the forester remains within the bounds of practicability, the stream that flows out of almost every forest will deliver the accrued interest, the timber, almost without cost. If the forester (already a forest-despoiler) continues to employ his present methods, altering the fundamental form in which forest flourishes (by clear-felling), then Nature will protect herself. The wholesale clearing of forest very quickly leads to the destruction of watercourses and to the ruination of a previously-profitable means of transport. The felling of a forest with limited access to water is no longer worthwhile, even if the distance to the water is slight, and the forest is thereby saved at humanity's expense.[74] Alternative systems of transport such as forest railways have seldom proved economic in the long run, because they require extremely large quantities of timber in order to bring in a return on

---

[73]The frequently-reported floods in the 1980s and 1990s in Bangladesh similarly derive from recent deforestation in Tibet, Nepal and north India, with catastrophic effects on the Ganges floodplain. - Ed.

[74]Viktor Schauberger was never opposed to the felling of timber as such, since he considered it an extremely useful natural commodity. His abiding concern, however, was that forestry and river regulation should be carried out in accordance with natural principles and therefore sustainably. - Ed.

capital outlay. In such cases the efforts of entrepreneurs to balance books through rigorous exploitation of forest has introduced such dangers to the national economy - that once a government becomes truly aware of the seriousness of the situation, it will have to enact the harshest of restrictive measures.

It is unthinkable that water should be controlled by mathematical formulae alone. The proper management of water above all demands a sense of commitment and great sensitivity, similar to that of a good doctor.

There are unmistakable symptoms in relation to the management of water which should once more be summarised briefly. As long as a waterway can transport timber unaided and therefore free of charge, the forester may use his axe. The deterioration of waterways warns of the dangers which, *without exaggeration, most seriously threaten our own existence.*

As long as the trout continues to stand motionless in the water because food flows unaided into its jaws, then favourable conditions will also exist for humanity and for the economy. If water exhibits destructive phenomena, if the trout becomes agitated and the timber begins to strand, then in the same ratio that the creatures in the water decline in quality, the conditions for all life will begin to disappear, conditions that those closely bound to their native soil can now no longer overlook. Contemporary methods of torrent-limitation, river regulation and hydro-electric power generation in general will have to be changed radically.

The increasing *karst* development in the upper reaches due to the continued sinking of the groundwater table, the destruction and devastation of the cultivable land in the lower reaches, the unruly and undisciplined discharge of catastrophic floodwaters into the valley, the increasing development of swamps in low-lying areas, the constant increase in the severity of so-called 'natural' disasters occurring locally from year to year, the decline of agriculture and so on, in many instances can also be put down to the totally unnatural systems applied to the regulation of rivers.

Large-scale regulatory works have been put in hand without even the vaguest idea of the energetic principles governing Nature's processes and without an inkling of the most fundamental laws of the movement of water. These have radically altered the natural scheme of things and acted in garing contravention of the laws here prevailing. Instead of bearing in mind the obvious fact that forest and high-altitude vegetation are just as essential as the skin on the body, everything possible has been done to destroy Nature's truly marvellous interdependencies, which otherwise are almost indestructible.

In the belief that forest exists to be exploited for every conceivable purpose, every effort has been made not only to plunder what Nature needs for life and for the maintenance of the soil as objects for vulgar speculation, but on top of everything, to destroy them through totally perverse practices. The most discouraging aspect is that in spite of all the bad experiences, these absurd methods of regulation and forest management are still followed today. Through such methods the forest, the prime necessity for culture of any kind, must demonstrably die as a result of measures presently applied by the responsible authorities. In no single instance can it be demonstrated that the regulation of even a small stream was carried out without fault.

Millions of people are already unemployed. Thousands of farms are in danger of collapse. Even very ingeniously-constructed machines are no longer able to work the exhausted soil in such a way that the energy expended is in proportion to the yield. Through devastation of forest and misguided regulation of our rivers the state of equilibrium in Nature's household has been disturbed. Such a crisis in agriculture and forestry, directly responsible for all other economic upsets, could never have occurred had forest and water been treated even half-way intelligently. In every way the hydraulic engineer has treated the waterways no better than the forester the forest. In view of the close connection between forest and water, it therefore comes as no surprise that channels are in an even more disproportionately dismal state than the forest.

Even at a time when thousands are destitute for lack of work, it is almost pointless to wait and see whether humanity, who has already lost all connection with Nature, will not only continue to destroy all further existence, but also shatter the very last hope of recovery. In reality, twice as many are required to arrest the total economic collapse that threatens us, which can only happen, if the mistakes that have been made so far are rectified as fast as humanly possible. This depends on whether the forest can be built up again as it once was and must always be, and if the channels are brought into balance once more through the construction of suitably designed reservoirs, so that at the very least they can keep to a tolerably regular course for the time being and generate healthy water again in order that healthy blood can once more be supplied to flora and fauna alike.

Tractive Force Considered
[From *Mensch und Technik,* Spec.ed. relating to rivers [Vol.2, 1993, Sec.8 Author's Notes from 7 July 1939]

The less the tractive force of a waterway, the more it lacks the inner cohering tensions that hold its body together. The result of these losses of tension is seen in the outfall and deposition of the sediment that furnishes a properly flowing stream with the provisions for its journey. Whatever these sedimentary substances have surrendered to their former creator in terms of their intrinsic valence, will once more be carefully deposited on the riverbank, there to be painstakingly decomposed by the Sun and in this way conducted back into the renewing cycle of life. In a healthy waterway the water-filaments migrate from the banks towards the centre and to the bed. Cradled in a normal profile, such rivers flow away tranquilly and crystal-clear, the teeming fish-life bearing witness to the superior quality of such water.

If the water's tractive force disappears due to the inner-atomic de-energising caused by incorrect profiling, senseless straightening or direct solar irradiation, then the depth of the sediment bed-load increases and every widening of the banks and flattening of the bed is followed by a deepening. As a result the insipid, warming water loses the last of its inner supportive substances. Becoming stale and tired, it winds itself here and drags itself there in order to find the substances that sustain its very existence. In this condition the *'expert Pied Pipers of Hamlin'* overpower the water and force it into narrow, straightened channels. The water flows through such truncations on a steeper path. Compressed between walls, it actually does flow faster and even transports its sediment conscientiously and diligently. This type of tractive force, however, is only very short-lived.

After a lengthy interval the flow-regime again changes drastically. Immediately below the regulatory works the mechanically accelerated water again begins to disgorge its sediment. As a result of this deposition, the channel inevitably widens again. Such widening can be observed after every narrowing and thus the evil described above is markedly aggravated. To counteract the effects of the widening of the channel, dams are built. This ultimately gives the water the possibility of annihilating everything at one fell swoop.

During the day the insolated water is a loose-knit, element-hungry entity into which the Sun impresses its substances. At night the temperature falls and with it the water's inner tension. Energetic essences of an immature, highly polarised nature[75] come into being as a result, which as manifestations of a less developed order of life, require new nutritive energies. They therefore steal from the soil surrounding the channel, taking those elements needed for their own sustenance and propagation. The discharging of the surrounding soil, both energetically and materially, and its nascent unwholesomeness is therefore the result of the improper regulation of the river.

---

[75]Described in the original German as *Differenzstoffe* (difference-substances). This refers to diverse elements, essences, or ethericities, whose mutual differences or natural characteristics are, or have been, intensified for one reason or another. In other words, they are in a highly polarised state, their various charges and potentials thereby approaching a more or less extreme monopolar condition, a condition where their differences are equally extreme, hence the expression 'difference-substances'. In this unbalanced state they develop an extreme attraction, or hunger, for their counterpart, a thirst which until assuaged, disables them from any creative or formative activity. - Ed.

Early in the morning when the Sun begins to shine again, fertilisation processes take place within and between these diverse essences and whatever is of high quality rises upwards and the next higher life-forms evolve in the water. These we describe as bacteria, which have come into being through the ethericitical fertilisation processes discussed previously. As long as sufficient supplementary material is still present in the riverbank, which as dynagens charge the water upstream for many kilometres, then viewed superficially, the pattern of flow will vary only slightly. This happens if the original direction of radiation has reversed through the change in polarity.

However, if the surrounding soil becomes impoverished, then the water, now no longer able to obtain its life-giving essences from the transverse potential-field, begins to appease its inner hunger by taking from the longitudinal axis. In this way the water extracts its life-energies from the direction lying closest to its source. In other words, it begins to discharge the upper reaches and thus, conditioned by the ensuing drop in tension, the deposition of sediment increases in those places that were previously regulated or where the sediment was forcibly transported a little further due to mechanical influences. Despite the re-supply of energy from upstream, tongue-like patterns of deposition form initially in the lower areas, which become broader and broader, ultimately silting up the channel entirely.

In earlier treatises it was clearly proven that in an average channel, the losses in tractive force arising from the inner de-energising of the waterbody, are in the order of 30-50 million horsepower, or 20-40 million kilowatts. It is common knowledge that in Nature nothing is ever lost, but always reappears in potentiated form. Every substance is bipolar and hence the ceaselessly recycling energies can be potentiated in either a beneficial or a destructive manner.

A watercourse that has been conventionally regulated gives no-one any peace and year by year the maintenance costs of such a channel increase. This well-known fact is ample proof that the character and quality of regulated rivers do not improve, but deteriorate. Now high and dry, the particles of sediment lie directly in the path of the scorching rays of the Sun. The metabolic processes generated in the stones by the alternation of night and day, lead eventually to their disintegration. The sedimentary particles that normally charge and discharge themselves from the ground upwards and from above downwards, invert their charges. In this fashion those substances that in the normal course of events the air receives indirectly via the water and the plants, now pass directly into the atmosphere. This results in a massive increase in the highly polarised essences, because the formative and distributive pathways so vital to Nature have been disrupted. Inferior and thus unsuited to the formation of the atmosphere, these undeveloped essences or substances of lower origin, i.e. the masses super-saturated by solar radiation - deploy themselves horizontally, breeding surreptitiously and wait for just the right moment to return to their terrestrial habitat or to recharge themselves on the Earth.

This fall to Earth will be caused through disturbances in the formation of the air, which also result in the excessive propulsion of the higher essences of life to even higher regions. The atmosphere's harmonious equilibrium having thus become unstable, de-energising phenomena now occur in the air. These, however, take effect in a manner completely different from those previously described in water. The de-energising of the air gives rise to the condensation of the remaining aeriform substances into which the essences of inferior provenance have now become incorporated. These return into the Earth as the water suddenly plummets earthwards. At this point something happens, which is crucially important for later developments and to which the strictest attention must be paid.

Coolness follows rain. This happens because the falling rain attracts those substances to itself, which were still too inferior for a later, higher outbirth. Should these substances be intrinsically beneficial, then they act on the Earth in a benevolently potentiated way. If they are unwholesome substances, however, then the potentiated unwholesomeness expresses itself in a rapidly proceeding deterioration.

To get some idea of the elemental forces that destroy a river and the water in improperly regulated channels, we must once again draw mathematical parallels and refer to the tractive forces mentioned earlier, so that we can construct a half-way useful conception. In a channel conducting about 500 cubic metres per second, about 60 million horsepower or 50 million kilowatts are lost, if it is warmed to about +22°C on hot days. These tractive energies are the dragging, suctional forces appearing in the whole waterbody. They endeavour to extract negative dynagens from the sediment so that, together with the phos-substances[76] implanted by the Sun, interactions resulting in the formation of new water can take place. These events are necessary so that the flattening gradients in the valley can be overcome through the increasing weight of the watermasses. This energetic discharge from the female substances of the sediment can only occur, however, with very specific vertical graduations in temperature. In other words, the relation between surface and bottom water must accord with a particular temperature gradient.[77]

---

[76] It is not clear what substances and properties Viktor Schauberger was referring with the term *phos-substances,* but in Phaidon *Concise Encyclopaedia of Science and Technology,* phosphorus *is referred to as being "of great biological importance". On the other hand phosphor is described as "a substance exhibiting luminescence, i.e. emitting light (or other electromagnetic radiation) on non-thermal stimulation"* and it also has the property of converting *"ultraviolet radiation into visible light."* - Ed.

[77] Here a positive temperature-gradient from water-surface to riverbed is inferred. - Ed.

Should the whole water-body become too warm - the potential differences in badly regulated channels thus being lost - then no heavy watermasses can sink into the depths. In consequence, no interactions and no formation of new water can take place. Water unsupported by increase and renewal slowly dies along its course and eventually the stream dries up. The counterpart to tractive energies are the hitherto unknown impact forces, which appear when large quantities of rain suddenly fill up a channel that has lain dry for a long period.

We shall now describe in greater detail the extraordinarily mysterious development of these impact forces to understand how they can be harnessed to drive organic machines. Through the rapid cooling of the atmosphere after a downpour or a hailstorm, the water cools faster than the surrounding ground. The outcome of this non-uniform cooling is the emergence of a horizontally propagated potential. In the rapidly cooling water the previously incorporated phos-substances become concentrated.

Under the sudden influence of cold, the female dynagens in the sediment begin to radiate their energies. Due to this reversal of inner tension, massive interactions take place in the opposing substances. Through these sudden inter-exchanges the creation of young water, or the strong precipitation of hydrogen occurs, which seeks to obtain its perfective or formative essences from the surrounding soil or the overlying air. A pressure potential evolves, which proceeds from the centre of the water sideways and upwards. This pressure potential, which acts at right angles to the direction of flow, impedes the draining water's normal axial motion. In the process transverse water-mountains form, which under certain conditions can reach a height of 80 cm. These forces are considerably greater than the heavy, mud-filled water flowing down a steep gradient.

Small fluctuations in temperature in the flow cross-section, which result from the cultivation or deforestation of the riverbanks, cause the axis of these transverse barrel-vortices to incline downwards. Highly charged with pressural energies, these heavy watermasses impact against the banks and in this dangerous state can even destroy metre-thick walls, because the wall's pores have become permeable to them and in the structure of the soil or the wall further interactions occur, whose effect is positively explosive. These explosions in water can be demonstrated experimentally and, since we are here concerned with high-tension, hydrolytic interactions, under certain circumstances act like liquid (compressed) air.

Tractive forces are the suctional forces evolving in the longitudinal axis. Pressural forces are the impact-forces developing in the transverse axis, whose potency is greater than the former. These laterally-operating, impeding forces normally prevent the over-rapid descent of the water on very steep gradients and represent the water's remote-controlled brake as it were. Up to now these hydraulic brakes have remained entirely unnoticed.

Similar processes in our own bodies are responsible for the circulation of the blood. The whole movement of sap is also to be attributed to the formation of tractive and impact forces, which function in a similar manner. The high wave-trains in southern seas after sunset, the incidence of heavy surf and all ebbing and flooding are also to be ascribed to these inner energetic interactions. In the final analysis the movement of the Gulf Stream is the product of these powerful impact forces, which come into being through a form of water-thunderstorm. Vertical water-squalls are initially triggered off, which then ebb away in any direction and have an effect as far as 1,000 km away. The formation of storms, hurricanes and notoriously dangerous cyclones in the atmosphere are to be attributed to similar causes. These potentially destructive elemental forces, which on occasion have been triggered involuntarily, may be harnessed to useful work in miniature machines in the future.

The greater the number and operation of such machines, the fewer the number of catastrophes that can arise. In the future Mother-Earth will again be able quietly to build up her energising substances in all peace, which today are so foolishly combusted. In Nature there is only a certain *'either - or'*, the *'I can live'* or the *'We can destroy'*. Outwardly the *'I-life'* is animated mass. The *'We-life'* are more highly evolved races that can affect the *'I-life'*-consciousness beneficially or detrimentally, depending on whether humanity correctly or incorrectly influences the stupendous synthesis, which is far from finished with the shaping of humanity. It would be absolutely wrong to say that after this life there is no later existence.

This after-life, however, is a phenomenon, which in its progress towards the Divine, congregates together in groups and no longer has any I-consciousness, but is absorbed by even higher powers through the agency of longitudinal and transverse potentials. It either directs itself upwards or downwards and annihilates everything, if, through senseless intrusions and blunders on Earth, a foolish mankind disturbs this *ur*-mighty course of evolution in order to exploit the forces that equip Nature, which are the foundation for the whole formation of life. To exonerate the nature-alienated experts responsible for this, on the grounds of ignorance and thus a certain innocence, does nothing for the hundreds of millions of their fellow human beings, who have to atone for their folly and forgo whatever little joy of life they have, merely because these so-called erudite have been so stupid.

Some will now say, *"Why not reveal the true knowledge immediately?"* There is only one answer for such short-sighted people: To make use of elemental forces in miniature machines is to place lightning in the hands of babes. Hitherto the ability to hurl thunderbolts was only given to highly evolved races. What would happen if the boorish masses of humanity, who presently squander all their vital energy in order to discover means to destroy one another, were to carry these gigantic *ur*-forces around in their pockets, so to speak. For this reason those who see things clearly carry an enormous responsibility and we must therefore proceed with extreme caution and indicate the way in which today's destructive forces can be transformed into useful energies.

Concerning Rivers and Water

Viktor Schauberger had a long correspondence with Dr Dagmar Sarkar in the early 1950s, during which time, having much confidence in her ability, he asked her to translate all his works into English. This she unfortunately had no time to do. Upon moving to India with her Indian diplomat husband, she did, however, continue to promote Schauberger's ideas by publishing a number of his articles and letters from their correspondence. I first became acquainted with Dr Sarkar in 1977 when she agreed to check the accuracy of my first translations. This contact was maintained over the intervening years during which a long correspondence and friendship developed. This passage of Viktor Schauberger's and one or two others were kindly made available to me by Dr Sarkar for inclusion in the overall translations. - [Editor]

The naturalesque conduction of flowing water is by no means a simple matter. It is an art demanding profound, empathetic insight and sensitive awareness. This is absolutely lacking in the hydraulicists concerned with the matter. Because of this there is hardly a waterway which is still healthy and which acts on the environment in an invigorating way. Hence, in the naturalesque conduction of water, the following should be taken into account:

1. the normal profile, which comes into being automatically through a biodynamic form of motion;
2. the rotation about its own axis or about its true, higher self, the `I`, which is a secondary result of a normal profile;
3. the purposive isolation through naturalesque deposition of the film of sediment;
4. the possibility for the water's up-welling, overarching curvature, which is enabled through vertical and horizontal curves. Plan and elevation are therefore one and the same, for only in this way can circumfluence take place. As a further consequence, an inner restructuring and qualitative enhancement thus comes about. The water is then endowed with the

power to build up its protective external skin, which ensures water's contrasting status in relation to oppositely-energised surroundings, through which metabolic processes can take place in all dimensions - in solid, liquid, gaseous, etheric and energetic states. Such water breathes, pulsates and is healthy. If it is unable to build up its sealing skin, then it dies like a human being whose skin has been burnt off.

Nothing could be more absurd, therefore, than to increase the geological gradient by channel-truncation, because in this case, owing to its self-weight, the water overruns itself. In the same way that a wrongly-ploughed field loses its soil-energy and soul-force (psyche), water also loses its inner nature if its maternalistic, uplifting (levitating) forces are missing. The same is also true if sediment is not dissolved by biodynamic processes and is unable to release the formative substances it contains.

The Sun can both build up and break down water. Every new entity can evolve only in its own amniotic fluid. Whoever would increase water must not only take note of the Sun-god's path, but also afford him the possibility of fertilising the water up-welling towards him in a naturalesque way.

The more intimately and aggressively the contra-directional, contra-potentiated substances interact with each other, the more active and cool the given substance becomes. However, if Man organises a watercourse the same thing happens as occurs in the forest, wherein true chaos comes into being through his mortifying order, out of which disease-causing [or pathogenic] bacteria evolve.

In every substance there is an innate drive to change itself in order to become mobile. If we move naturalesquely, then through the metamorphosis of substance or through the inauguration of a *naturalesque* metabolism, we achieve the desired velocity or raise the efficiency. In such an instance we are operating economically and in a fuel-saving fashion. If we move unnaturally, then we achieve the opposite. The regulation of a watercourse - bank rectification - results in the internal decay of water's substance and in its disappearance - which we can preserve through its organisation.

Every change in a substance acts to create a resistance which impedes the otherwise-unbridled intensification of the will-to-change. Without this resistance there would be no sustained movement and no animation or enlivenment. This self-increasing resistance actually becomes the life-force itself. For this reason the product of the concentration of matter only has any value if it has undergone the highest qualitative ennoblement or the profoundest debasement, either naturally or artificially. Only in this way can kinetic energy be raised. The longer the developmental path or process, the shorter the developmental period becomes. Conversely, a purposive shortening of the reversed developmental path or the shortening of the reconstituting period signifies an acceleration of movement and an increase in kinetic energy. A purposive organisation of the aforementioned metabolic forces, however, enables the elimination of resistance at virtually no cost. In this case motional resistance decreases by the square of the velocity, whereby unimaginable outputs can be achieved with the smallest input of energy.

Today's technology has not investigated the will-to-change in all matter, but has built a motion-inhibiting resistance into all machines.

The Transport of Sediment
Fundamental Principles for the Floatation of Timber, Ore and other Materials Heavier than Water
[An excerpt from *Tau* 137, pp12-15.]

1. Water is no lifeless substance, but an organism. It is the blood of the Earth, subjected to an uninterrupted interchange of substance (metabolism), causing a constantly-changing pattern of movement. Water is the carrier of substances which condition life. All we see around us has been built up by powerful forces that dwell in water. Once we understand the fundamental law governing this gigantic developmental process, then we shall also be in a position to shape all we see around us creatively, in the way that Nature shows us. Everything flows, floats and moves. There is no state of equilibrium - there is no state of rest. If we want to live, then we must become initiates and understand the universe and its inner conformities with natural law. We must bow down before Nature, who has given all of us, and will continue to give us what we see, feel and hear, once we take heed of that which gives us life: the dynagens inhering in water.

2. The product of these metabolic processes is motion. Tractive force and rate of flow are conditioned by the temperature gradient. The temperature gradient is thus the regulator of movement.

3. A particular profile is contingent on a particular temperature distribution, which in turn determines the molecular distribution of matter. Conversely, a particular molecular distribution determines a particular temperature distribution, which on its part determines the profile. The development of the profile and the disposition of the river-bends is thus a question of the organisation of the various water temperatures. Water temperatures are opposites, and these opposites condition life and determine the type of motion.

4. The composition of channel wall-surfaces determines the coefficients of friction and the distribution of heat and cold as actual substance. Rough wall-surfaces or inbuilt rifling increase frictional surfaces and the incidence

of heat. Where heating occurs, tractive force diminishes. Reduction in tractive force means deposition of sediment - stranding - and therefore the formation of the riverbanks. Through the withdrawal of heated matter from the water, cooling occurs in the flow-axis, and with this a raising of tractive force in the same. The core water-masses forge ahead and deepen the centre of the channel. Advancing water sinks vertically along the axis and suspended matter is conducted down the centre, because it is here that the channel is being deepened.

Suction forces evolve in the core-water body, because here the translatory velocity is greater relative to the peripheral flows. This phenomenon results in reduction of pressure on the riverbank. Due to these centripetal forces, the walls of the riverbanks are relieved of pressure to such an extent that the walls merely serve to render the banks watertight and impermeable. This impermeability is achieved through sedimentation (the deposition of the finest particles of sediment - silt). This is a necessity, because in this way a vital separation is maintained between surface water and groundwater.

5. The more specifically dense and more homogeneous the core water-masses, the more buoyant the suspended matter becomes.[78] The greater the volume and velocity of the core water-masses, the more powerfully the centripetal forces act to draw the peripheral flows towards the central axis of the current. The outcome of this is a progressive reduction in pressure on the riverbanks, an increase in the tractive force of the core water-masses, an increase in the mean overall depth of the water table (groundwater recharge) and an increase in buoyancy of the core water-masses, whose homogeneity increases with depth - they become more laminar.

In this way suction is created with a commensurate increase in tractive force, causing those particles of sediment heavier than water and of a particular grain-size to be sucked into the specifically-heaviest core-zone. There the greatest density and hence the greatest flow-velocity prevail, and thus the resistance to flow is also here at a minimum.

Bodies heavier than the core water-masses will be carried down the central axis through the combined effect of centripetal, negative pressure and translatory velocity. Because it moves more slowly, sediment (sand, dyestuffs, dirt) lighter than water will be ejected laterally and progressively deposited according to weight. The closer to the bank, the lighter the substances deposited.

6. Heavy, centrally-conducted particles of sediment will be reduced in size mechanically through their mutual abrasion. The stored energies in these energy concentrations will be released and subsequently absorbed into cold core water-masses.[79] In this way the status of water is enhanced biologically,

---

[78] Water density and weight equals or is greater than that of the suspended matter. - Ed.
[79] See footnote 4- Ed.

which process extends to the vitalisation and stimulation of the groundwater body, which now has to move on its part - and it rises. The fertility of the soil increases.

7. River regulation with the aid of the temperature gradient signifies:

- the possibility of transporting bodies, whether lighter or heavier than water, up to 90% more cheaply than hitherto has been the case with all other systems of transport, by exploiting the natural rhythms of motion (normal profile);
- an automatic development of a normal profile, and
- the raising of the groundwater table and an increase in soil productivity.

8. Contemporary systems of river regulation cause:

- loss of the cheapest means of transport;
- the deterioration of the waterways;
- the sinking of the groundwater table and a decline in soil fertility.

9. Accordingly, current systems of river regulation (temperature gradient disregarded) lead to the aggravation of the crisis. This leads to pollution of the channel, disappearance of water and an increase in diseases - economic decline.

10. Future systems of river regulation (temperature gradient considered) consequently help to overcome the crisis and the rehabilitation of the whole. It leads to decontamination of channels, increase in water through promotion of the necessary interaction between opposites, and to economic growth.

No water, no life - Bad water, bad life - Good water, good life.

Current statutory hydraulic regulations are a key contributing factor towards decay and the progressive spread of cancer.

Viktor Schauberger, Vienna, June 1935.

The Rhine and the Danube

The Problem of the Danube Regulation
[From *Implosion* Magazine No. 23.]

Viktor Schauberger conducted a long battle with the scientific and government institutions of his day to save the Rhine and the Danube from total ruin but it was a battle which was ultimately lost through their rejection of his practical suggestions. In early 1932 he wrote a paper about the rehabilitation of the Danube, detailing the measures that needed to be taken in order to reinstate the once-magnificent river to its former glory. This paper was included in The Danube, a study undertaken by the International Danube Commission. Upon discovering Viktor's contribution to this major work, however, officials decided, at great cost. to recall and destroy the whole edition. When it was republished in October of that year, it was without Viktor's offending article. All this happened largely due to the actions of Viktor Schauberger's implacable antagonist, Dr Ehrenberger, who hounded him wherever he went. - [Editor]

The enormous, general and economic importance of the Danube, already eulogised in the *Song of the Nibelungs*, is fully established by the fact that every country wishes to have access to it. A special International Danube Commission has been established which is concerned with all related technical and economic questions. Although the most highly qualified experts are engaged in the study of all the hydraulic engineering problems, year after year the daily newspapers report on the catastrophes occurring during times of flood, which lead to particularly grave consequences, for the very reason that the protective measures set in place by man have failed.

Up to now attempts have always been made to increase the efficiency and thus the stability of protective structures by strengthening them. These constructions - the result of taking only the mechanical factors of hydraulic engineering into account, while completely neglecting the equally important physical aspect of the processes - have not only *not* fulfilled their purposes, but in the course of time have also caused inestimable damage.

Thus through the regulatory works in the upper and middle reaches of the Danube, 950,000 hectares of productive land were transformed into a flood plain, thereby being rendered useless for agriculture. Even in Yugoslavia (Serbia) whole villages had to be evacuated and their inhabitants relocated at the state's expense. Professor Vidrasku of Budapest wrote about this in the *Neue Freie Presse* (New Free Press) as follows:

"*Unfortunately, in the belief that construction of huge dams on the banks of the Danube would suffice, works were begun without having studied the problem in sufficient detail.*" He continues: "*If we were to eliminate flooding with massive, high dams, and no longer permit inundation of the flood plain, then the floodwater would rise so high that all our river-ports and all riparian settlements along the Danube would be under water. Moreover, river and lake navigation, which even today is unsatisfactory, would be significantly impaired, and the dams themselves would offer no additional security.*"

A further excellent and informative example relates to the raising of the riverbed of the Rhine at Salez. The valley cross-section shows that the Rhine flows along an elevated plateau, its bed lying up to 4m (13ft), and its flood-peak of 1890, 7-8m (23-26ft) above the lowest point of the valley, ultimately intersecting the rooftops of the villages in the low-lying areas. With good reason Mr Otto Rappolt, a chief government surveyor, states in his book *River Engineering* (*Flussbau*, published by Göschen Sammlung) that:

"*The deforestation of the catchment area and the system of river rectification is to be considered the principal cause of this hazardous predicament. Thus every important river in Europe provides us with sufficient cause to consider the consequences that incorrect river regulation can draw in its wake. The rivers Etsch, Po and Tagliamento in particular will present the Italian Government with many more problems as well as swallowing up considerable amounts of capital, if a radical departure in present government policy is **not made**.*"

Thus every major river-system provides us with examples of the consequences resulting from faulty river engineering. In order now to arrive at the right results, it is necessary to augment the mechanical aspects of the process with physical influences, of which the most crucial are the temperature gradient, the variation in specific weight and the occurrence of material metamorphoses. The enormous significance attaching to the temperature gradient is best perceived from the fact that to heat up $1m^3$ of water by only 0.1°C - and such variations in water temperature are to be found in *almost every* channel cross-section - roughly 42,700kgm are required. From this it is evident what tremendous energies are freed with a reduction in temperature or bound with an increase in the same.

Both *'positive'* and *'negative'* forms of water movement, which hitherto have been referred to by their outer appearance and described as 'laminar' or 'turbulent' flows, represent unknown magnitudes in the equation to be solved. From this equation are to be derived the great fundamental laws of formation and growth, which not only enable autonomous development, but also the degeneration of all forms of vegetation and the here equally important laws of destruction.

Under the term *temperature gradient* is to be understood the variation in water temperature during its movement. Water, whose temperature gradient approaches +4°C (whether or not this approach is from high or almost-freezing temperatures is immaterial), we can consider as being under the influence of a *positive* temperature gradient. Water whose temperature diverges from +4°C we describe as being under a *negative t*emperature gradient. At the ground-surface, under the influence of the atmosphere, water in a positive temperature gradient will accumulate its oxygen groups while its carbone groups will be uniformly distributed within it. Also its flow will converge, its bed will be hollowed out into a semi-circle, its tractive force will increase and its flow will accelerate - in other words, it will assume a straight-line motion. Conversely, if water flows under a negative temperature gradient it disperses its oxygen groups, accumulates its carbones, absorbs large quantities of atmospheric air and in the same ratio loses its tractive force, deposits its transported matter, gravel, sand and dissolved gaseous carbones. In the organism of the Earth, water is therefore a carrier of oxygen and carbone groups, and at the same time it is a carrier of all the organic processes which act to build up all life in Nature. The exchange between these two groups of substances is not only of crucial importance for the formation of all organisms, but also for the behaviour and characteristics of water itself.

Proceeding from the basic premise that surface and groundwater flows are actually *arteries* whose sole purpose is to convey various substances to the environment, we arrive at the following perception. If a river is badly regulated, it not only results in frequent sinking of the groundwater table, thus affecting the vegetation, but the vegetation itself will also be seriously afflicted due to physical causes. If, in relation to its surroundings, such water is under a negative temperature gradient - comparatively warmer oxygen-rich water, hungry for carbones, is present in the channel - it will extract carbones from a wide area and instead of being impressed into the root-zone, these carbones will be drawn towards the greater attracting force - the channel. This results in constant and progressive decline of the vegetation in such areas. Instead of going into the ground, the water enters the widening channels and as a result of too high a temperature, it is transferred to the atmosphere too rapidly. In this process not only is the full hydrological cycle unfavourably affected, but dangerous flood catastrophes must inevitably follow.

In future stream and river regulation we have finally to abandon the contemporary perverted, materialistic way of looking at things, which views water merely as an element destined to drive numerous power-stations along its course, to be the recipient of waste water and to be conducted to the sea by the shortest possible route by way of a whole series of hydraulic paraphernalia. Water's appointed tasks in Nature's housekeeping, which are to a large extent still insufficiently evaluated, must once more be taken into account. The large measure of purposefulness in natural life and water's vital role must be properly understood once and for all, and attempts should no longer be made to control water with extremely expensive, totally irrational mechanical means of coercion.

In the following the grossly discounted *physical processes of metabolism* will be addressed - this metabolism takes place between *oxygen and carbones groups* and *influences* the development of the channel. These processes take place on a very small scale, even though in infinite succession. They are those pulsations which can be detected experimentally through the fluctuations in the water level in a Darcy-pipe or through the continuous fluctuations in the rotational velocity of Woltmann-vanes.[80] The oxygen-deficient water emerging from a spring absorbs oxygen from the air, which sinks to the bottom due to its molecular weight, whereas the water's content of carbon-dioxide will be evenly dispersed by reason of the cold water's high capacity to absorb gases. Because of the warmer temperature-influences of the ground and the atmosphere active along its further course, spirally-disposed isotherms appear in the water. In the presence of aggressive oxygen (O^3, ozone) in the oxygenated zone, the close packing of these isotherms leads to cold oxidations, through which various highly complex carbone groups are formed under explosion-like events. The most highly-organised substances will subsequently be impelled in part towards the riverbed and in part towards the banks, where they serve as raw material for the growth of the adjacent vegetation. Less-refined substances, during whose formation (explosions) minor energies only are released, are borne downstream under the influence of the current. Apart from this, a further effect of the explosions described above is to propel smaller suspended matter transported by the water from one epicentre of explosion to the next and ultimately towards the riverbank. It is said that the water 'runs aground'.

If the influence of the warm temperature persists, then water's content of carbonic acid escapes in familiar fashion through the formation of carbon-dioxide bubbles, their place now being taken by oxygen, which distributes itself uniformly through the water. Since this warmer water also has for the time being less tractive force, suspended matter sinks to the bottom. The water therefore finds itself in a depositing mode through the above-described explosions and also as a result of the reduction in tractive force engendered by the rise in temperature.

[80] A Woltmann vane is hydrodynamically shaped device suspended in streams to measure the variation in the velocity of flow. - Ed.

The water now begins to extract fresh carbones from its bed and banks, which cool it. Since the carbones now lie at the bottom and the oxygen is evenly diffused in the water above, explosions will now occur in a direction opposite to the one previously described, which act in a 'shovelling' way. The water *excavates*. The excavating phase is then followed by a *depositing* phase. The carbones present in the deeper part of the channel profile become dispersed as they rise towards the surface and the oxygen sinks towards the bottom, reintroducing the depositional phase. In this closed cycle of alternating deposition and removal, the quality of the water continually deteriorates and it becomes 'sluggish and stale, lifeless and insipid'. This latter phenomenon has two main causes:

(1) increase in water temperature and the associated decrease in specific weight, and

(2) the fact that the water available at the beginning of each phase is always of poorer quality, because substances least ennobled by the explosions are driven downstream where they become the starting product for new explosions.

As long as healthy conditions in the water prevail - as long as the proper proportions between the inner water-masses, the bed-gradient and the temperature continue to exist - the incidence of turbulent phenomena will be stronger along the upper margins of the banks while constantly weakening towards the bottom, and the axis of the current will remain unchanged. The constantly-increasing strength of the external influence resulting from the widening of the channel reduces the water's tractive force. Towards evening the temperature gradient once more approaches a positive energy-form and during the night sediment will be transported further. At a suitable temperature the watercourse regulates itself automatically in every respect. With its temperature properly adjusted, the watercourse will not only widen itself in the proper proportion, but will also deepen itself appropriately in the lower reaches. Thus the longitudinal profile suited to the varying cross-section is formed automatically - a necessary matter for the removal of the sediment load, which under orderly forest conditions is in any case minimal. Where undisturbed by human hand, when flooding does occur in such waterways, the influence exerted from the outside on the increasing water-masses is stronger, due to the external temperature which as a rule is low at such times. The temperature gradient becomes positive and the now cooler water-masses drain at a greater velocity *without attacking the banks*.

With the correct reversal of temperature, the influence of temperature gradient is greater than that of bed-gradient, because the mass-transport of sediment will be appropriately regulated in proportion to the corresponding increase in flow-velocity. In this case, with an increase in the volume of water, the speed of forward motion will also be greater due to the proper rearrangement of the temperature gradient. The increase in energy required to dislodge and transport the sediment in the lower reaches is achieved through the introduction of cold water from shorter affluent streams, with which the water temperature of the main channel becomes reduced. Even Weyrauch, in his book *Hydraulic Calculation*, admits that without affluent streams the boundary condition of the tractive force decreases. In view of the above explanation, this phenomenon would appear to be quite self-evident.

From a mechanical point of view, according to Poseille's equation, the associated increase in *internal viscosity increases* by the square with a *decrease* in temperature, in the equation

$$P = 16\pi\, Y\, av$$

where P remains constant due to the increase in the magnitude of Y, a lesser velocity v would already suffice to maintain the state of equilibrium. However, since v will not be reduced as the bed-gradient is constant, the horizontal component a in the interplay of forces involved in the transport of sediment will be all the more active and hence the tractive force will be greater.

It should be mentioned here in passing that rivers flowing into the sea under a positive temperature gradient (*eg.* into the Arctic Ocean) carry their sediment far out into the sea (haff formation), whereas rivers debouching into the ocean under a negative temperature gradient (*e.g.* the Nile) deposit their sediment prior to the confluence (delta formation). At this juncture it is very interesting to note that according to Dr Schoklitz's laboratory experiments at the Technical University of Brünn, the law stating that the sediment-load increases in proportion to the square of the depth holds true for grain-sizes of up to 4mm (0.16in). In this regard, at the Research Institute in Vienna, Professor Schaffernak has discovered a law which holds that the sediment load is 'directly proportional' to the increase in depth for grain-sizes of 100mm - 130mm (4-5in) commonly found in our rivers.

Apart from this, we also discover a splendid array of inconsistencies amongst leading authors concerning the mechanical measures applied to hydraulic engineering as well as their mathematical treatment. Here attention is to be drawn briefly to the extremely varied opinions concerning the value of the great Kutter formula, to Tolkmitts' and Lindboes' critique of the Bazin formula, to the discussion about the coefficients in Christen's velocity formula, and others. It is left to the arbitrary discretion of each individual to increase or decrease the mathematical accuracy through this or that formula convenient to him. Nature works uncommonly slowly. This is why it is also impossible to encompass the great processes which take place in Nature by means of laboratory experiments, because the proper preconditions and interrelations are lacking. For this reason, in river engineering the causes of the resultant effects should only be studied in Nature's great examples. Observation over many decades is necessary in order to understand truly the infinitely fine, constantly increasing interplay of forces which are visible to the eye only by their mechanical effects, whereas the causes themselves mostly remain unnoticed and unheeded.

The treatment of water by mathematical formulae alone is unthinkable. The proper handling of water requires great interest and as the hydraulicist Robert Weyrauch stated so succinctly:

"In order to carry out river engineering projects, an especial gift for hydraulics, an exceptional feel for what is hydraulically possible or impossible is necessary. This is only acquired with difficulty and even the most experienced repeatedly suffer disappointments."

We are reminded that those people who continue to believe that they can solve the difficult problems of river engineering and regulation by methods other than those indicated by Nature, have taken upon themselves a truly onerous responsibility and liability.

The Rhine Battle
[From *TAU* No. 142, p. 6. An open letter to the Austrian Prime Minister.]

Quite a while ago, Mr Werner Zimmermann outlined to me the catastrophes which are to be anticipated in the Rhinelands (St Galler, Rheintal) and which the daily newspapers are already discussing. The former Chancellor, Dr Ender, also described these perils to me in a similar fashion - perils which will happen if the overall level of the Rhine cannot be lowered by a few metres.

It is, and will be, impossible for this lowering to be successfully achieved through bank-rectification and dredging. The dangers will inevitably be multiplied, precisely because these measures will reduce the tractive force, which is here the critical factor.

This can only be remedied by *the organisation of the necessary tractive forces*. Sufficient proof of their absence is evinced by the constant increase in the looming danger due to canalisation, truncation, and bank-correction. At the time, and with the support of the respected hydrologist Dr Forchheimer and the world-renown scientist, his excellency Professor Wilhelm Exner, I fully intended to explain the reasons for this heightened danger. Unfortunately these treatises were suppressed by certain bureaucratic institutions and because of this I was also unable publicly to refute the absolutely absurd explanations put forward by the Minister's river-engineering adviser, Dr Ehrenberger. Due to the passing away of the Federal Chancellor, Dr Dollfuß, who repeatedly sought me out personally in this regard, I had no chance to explain the errors in Ehrenberger's report to him verbally.

Prime Minister, please accept that the following explanations have no financial motive, but are founded purely on humanitarian grounds.

As a result of all these personal attacks, I renounce all pecuniary advantage and for this reason the following explanations are intended to rehabilitate the reputations of two people who have my highest respect and whose purely humane endeavours I cannot permit to be negated by arguments lacking in material facts.

All motion of whatever kind is associated with an expenditure (loss) of energy, if on its moving path no reinvigoration or nourishment of what is being moved takes place. Water conducted along a uniformly-profiled channel is warmed by direct solar radiation or friction. Because of this warming, a slow but constantly active discharge of the essential intrinsic energies of atoms occurs, which results in losses in kinetic energy. These gradually increase and produce symptoms of fatigue, provoking the deposition of sediment and causing the upward displacement of the riverbed.

In order to heat 1 cubic metre of water by only 0.1°C (such variations in temperature are to be found in the smallest channel cross-sections), work or an expenditure of energy in the order of 42,700 kgm is required. If 0.1°C = 42,700 kgm, then 1°C = 427,000 kgm and 20°C = 8,540,000 kgm or 114,000 hp/sec. The Rhine conducts about 500m^3 of water per second, hence with a warming of 20°C, the total loss in hydraulic efficiency = 57,000,000 hp or 42,000,000 kW. Every 0.1°C increase in heat, therefore, signifies a considerable loss of energy, which must be replaced if extremely hazardous depositions of sediment are to be avoided. Dams are ineffective in this regard, because one day the upstream gradient eventually flattens out through the continuous deposition of sediment caused by slowing water-flow, and the water therefore has to break out sideways. With this a flood catastrophe is already at hand.

How can this flattening of the gradient be prevented? Only by the replacement of the necessary energies lost due to solar radiation. If we consider the energy-loss ensuing from a rise in temperature of about 20°C, then it is also clear how vital it is to replace these losses in *tractive force* systematically. Through the strong aeration occurring with acceleration (truncation, canalisation *etc.*) this loss is only exacerbated and for this reason other means must be applied in order to maintain the water's *tractive force*. These means are precisely those hitherto-unknown *pulsation phenomena,* which trigger off *cooling effects.*

Rivers which cool off along their course carry their sediment far out into the sea due to the maintenance of their kinetic energies (haff-formation). Conversely, rivers which warm up along their course lose these energies, form deltas and shift their point of confluence.

The great law with which Nature controls and directs flowing water has so far escaped science's notice. In water, two different processes of dissolution take place: (a) dissolution due to heat and the influence of light, and (b) dissolution due to cold and the absence of light and air.

In case (a), matter transported by water in suspended form is insoluble. The release of energies contained in these substances is dependent on the occurrence of certain processes of ionisation, which we know of from the `photo-electric effect`. Positive or positively-charged metals allow negative electrons to escape under conditions of heat and light. With negative or negatively-charged metals the process is reversed under the exclusion of light and the influence of cold. Both forms of radiation are measurable and have a magnitude of about 2 volts, if groups of heteropolar metals[81] are present.

Metals are found in all waterways. On account of their weight, the specifically-heaviest metals are first to sink to the bottom and are left lying there. The specifically-lighter parent rock from which they have been abraded travels further on. Thus in the lower reaches of a river a deficiency of metals and hence symptoms of fatigue (losses in *tractive force*) must take place, because no inner-atomic energies can be released for lack of ionisation.

Freed energies can only be bound in the absence of light and as the water cools. If the water is actually warm, then spallating energies (electrons) are necessarily lost to the atmosphere. In order that water can become internally charged, every process of absorption must be followed by further reactions which can occur only if the temperature gradient is maintained at a certain low level of coolness. These after-reactions are phenomena of electrochemical fusion. The gases are transformed into electrozoic (animalistic or organismic) energy-products. If this mutation cannot take place, then even the strongest ionisations are of no avail. If the temperature gradient is correct, then flowing water becomes an animalistic accumulator. Water which is too warm is acidiferous and electrically over-conductive, and because of this an inner discharge takes place which fatigues the substance itself.

Dredging only *aggravates* the evil, because the heaviest (metalliferous) stones sink into the dredged holes and in the main are lost. Through the emissions of inner-atomic energies, magnetic forces evolve which reduce the absolute weight of the sediment. As a result it behaves in a manner contrary to Archimedes' principle - the heavier (more metalliferous) it is, the more buoyant it becomes.

My system of regulation, or more accurately, my organisation of the water, results in the free floatation of stones or logs heavier than water down the axis of the river. This is because there the greatest cold prevails and the strongest processes of mutation occur, which render the water *gas-free*, dense and therefore *mobile*.[82]

[81]*heteropolar* = of a molecule or compound being or having a molecule in which there is an uneven distribution of electrons and thus a permanent dipole moment. - Ed.

[81]"Despite its slow flow, good, mature water draining down a steady gradient has a greater carrying capacity and tractive force than water with a high gas content." V.S. in a letter to Dr. Dagmar Sarkar, 1949

If water, whose forward movement is unimpeded, shoots ahead down the centre of the channel, then the *normal* or *double profile*, which we can observe in all natural watercourses, develops automatically. The blood and sap capillaries are also built up in this way, because without this double profile there would be no *manifestations of valency* (external charge), all of which are of a dualistic nature (positive and negative).

Herein lies the secret, sketched in broad outline, why in all civilised countries the waterways are being ruined and are attracting increased maintenance costs annually. Herein also lies the secret of the degeneration and death of the forest and the degradation of the soil. In a word, this is the cause of the *world crisis*, which is simply the after-effect of the *disturbance of the valencies*. The contemporary forester (clear-felling), hydraulic engineer (bank rectification), landowner (artificial fertilisation) and energy technologist (the pillager of coal destined for the *motion-of-creation* and not for the *creation-of-motion*) have caused this through total ignorance of the most elementary laws of Nature.

Energy-Bodies
[From *TAU* No. 137, p.19. - Werner Zimmermann]

Let us consider our problem child once more, the Rhine. Can one of three possibilities be of any assistance here? The first two: reafforestation and the resultant re-creation of cold affluent streams - these take too long and are too expensive to be able to satisfy immediate requirements. And the third: impounded lakes? Here too, in the case of the Rhine it would take a great deal of time and money before it could be realised practically. Is there a fourth possibility? The installation of energy-generating bodies in the axis of the surging water-body - and hence the cooling of the water. Here all the elements are lacking which enable us to assess how and whether such measures can achieve the desired goal. It concerns the discoveries and inventions of our researcher - Viktor Schauberger - and we must be satisfied once again with indications and hints.

[From *TAU* No. 137, p. 27. - Viktor Schauberger]

The lowering of the bed of the Rhine by 4-6m (13-20ft) is simply a question of the *status of the tractive force*. This can only be solved by regulating water *temperature* and costs only a small fraction of what conventional systems of river regulation usually devour.

Dredging is an absolutely absurd procedure. One flood suffices to fill the dredged holes up again. Indeed it should be remembered that the amount of sediment brought down into the valley by the Rhine annually is estimated at about 100,000m³ (3,531,000 cu ft)! Every increase in the height of levées, however, only increases the danger of a breach, which is *inevitable* should a *warm* flood one day eventuate.

They should commission me! Minimal expenditure will suffice to banish the danger forever. The successful regulation will be guaranteed. *The regulatory works will need only be paid for once the bed of the Rhine has sunk by about 2m (6ft6in).*

[From *TAU* No. 142, p.15. - Viktor Schauberger]

As far as I am concerned, just what the study was all about is as follows. When I *secretly* installed these energising bodies in the Steyrling stream about 14 years ago, within the space of one night the river was washed out to such a degree that hundreds of cubic metres of sand and gravel were thrown up in great heaps into the so-called sand-trap. The stream sank right down to the bedrock overnight.[83]

Just imagine what would happen if these energy-bodies were put to work in the *middle* reaches of the Rhine! Thousands of cubic metres of sediment would then be set in motion, and if great care was not taken, then the dams would also collapse. At the limit of the effective action of these energy-bodies, the Rhine would break out sideways and inundate everything. Should this be undertaken prior to the confluence, then the Rhine would be lowered by about 6 - 8 metres and hundreds of thousands of cubic metres of sand and gravel would thoroughly fill the Bodensee (Lake Constance) at the confluence.

In this situation well-considered measures and extreme care are the only solution. The Rhine must first be *repaired* and then its flow *organised*. This repair-work extends across three countries - Germany, Switzerland and Austria. Recuperating water actually begins to grow of its own accord (proof of this is provided by the Neuberg installation). For this reason torrent confinements should not merely be altered to prevent the descent of more sediment, but steps must also be taken to cope with the drainage conditions downstream.

Why are these powerful effects produced by the installation of energy-bodies? The water's oxygen becomes *fluid* (a process similar to the production of liquid air), resulting in quite enormous metabolic activity in the water-body, which now attains its highest level of energy. The water becomes crystal clear, fast-flowing and *cold*. The fish population increases because the water becomes energy-rich and healthy.

As it flows along its course, the Rhine loses about *half a million* horsepow-

---

[83] "*If water is additionally stimulated through the introduction of metallic bodies, an increase in the tractive force ensues with the result that stones and ores float like softwood.*" V.S. in a letter to Dr. Dagmar Sarkar, 1949

er through the generation of *heat. The heat organises itself at the expense of motion*, which has its origin in radiation. The generation of heat *shapes* and *brakes*. The development of cold is *motion-creating*[84] and acts to release rays of negative energy. Through my measures the character of the Rhine would change and it would become like a fast-flowing mountain river.

If nothing is done (the effects of dredging only *accelerate* the havoc), then one day the Rhine must break out and create a new bed, because in its old *sick-bed* it can no longer *move*. Perhaps you now understand the jittery anxiety of today's river engineers and what they are liable for, once people experience the extent of the mistakes that have been made. There can only be *one* choice for courageous, upstanding men: to own up to their mistakes honestly (for to err is human) and to assist personally in the work of rectification and reconstruction.

[From *TAU* No. 146, p.29. - Viktor Schauberger]

When they want to make ice cream, cooks throw salt into ice-water, by which heat is bound. The water freezes if it is stirred at the same time. This is not to suggest that I advise that salt should be thrown into the Rhine, but that my advice is to not dredge the stones out of the Rhine. These stones contain salts which, under the right preconditions, cool the Rhine and increase its *tractive force*.

If appropriate metals are caused to interact correctly with any given acid (such as water warmed and strongly oxygenated by the Sun) and their opposing charges arc (viz. Volta's element), then a juvenile gas is released which is usually described as hydrogen gas.

This evolving hydrogen gas absorbs the oxygen instilled into the water by the Sun and produces an inner vacuum. In this way a loss of heat follows, because water is actually created out of this material transformation - water is just as unable to grow as any other vegetation without the influence of heat. From the combination of (blue) water-gas, oxygen and heat, an intermediate substance or gas evolves, normally called either carbonic acid or carbon-dioxide. Since this newly-created product cools through the binding of heat, then for every 1°C cooled the new gas-product loses one 273rd part of its volume. The water becomes free of voids, homogeneous and dense, and in this way extremely mobile (mainly in the flow-axis). The outcome of this is an intensified material metamorphosis and an increased growth of the water, which arises as a result of the previously-described transformations (mass-increase). We are here concerned with the fact that the forces arising from this hydrolytic transformation are greater than those required to trigger it off.

---

[84]This assertion seems to pre-empt the discovery of superconduction, where resistance-induced energy losses decrease with increasing cold. - Ed.

Because of this, and despite the loss of thoroughly leached sediment and reduction of the mechanical bed-gradient, the magnitude of the water's kinetic energies will increase constantly. Therefore, regardless of the diminishing gradient, water is able to flow steadily with increasing volume and thus it is able to transport its residual sediment or its digestive waste-matter into the sea.

[From *TAU* No. 145, p.17. - Viktor Schauberger]

I could only prove this with the aid of particular *water temperatures*, because the *tractive force* increases in the flow-axis when organic hydrogen evolves through the proper mixture of temperatures and appropriate metal facings (with silver as catalyst). Silver binds the oxygen gases contained in the water and releases carbonic acid and carbon-dioxide from the sediment. This produces an extremely powerful water which is even able to transport stones and iron.[85]

---

[85] These egg-shaped energy-enhancing bodies may perhaps be alloyed with alternating bands of oppositely-charged metals such as zinc (-) and copper (+), or gold (+) and silver (-). - Ed.

The Dr. Ehrenberger Affair

[From *TAU* No. 142, pp.13-14. - Werner Zimmermann]
Is *Schauberger to be taken seriously?* This crucial question can be answered briefly. Mr Nesper, an engineer, refers to the fact that in *Die Wasserwirtschaft* the Minister's river-engineering adviser, Dr Ehrenberger, is supposed to have refuted Schauberger's assertions. However, in his answer Nesper first of all refuted his colleague, engineer Bühi, who in his report made me Schauberger's `mouthpiece` and tried to discredit Schauberger's statements as products of fantasy and me as a dreamer.

On the other hand, what of the 'Ehrenberger rebuttal'? In Vienna Schauberger showed me a whole sheaf of original letters addressed to the editor of *Die Wasserwirtschaft* in 1931, in which threats were made by influential public officials to the effect that they would cancel their subscriptions and ruin *Die Wasserwirtschaft,* if it ever published a paper or a reply of Schauberger's again. This is the true nature of *"the rebuttal of Schauberger by experts and reputed scientists"* .[86]

A second piece of evidence, which demonstrates just how these specialist scientists work, is presently in my hands and I willingly make it available for your perusal. It concerns the great work, *The Danube: its Economic and Cultural Function in Central and Eastern Europe,* published in October 1932 under the joint auspices of the International Danube Commission and the governments of the Danube States.[87] This well-illustrated *magnum opus* originally contained a paper by Viktor Schauberger on page 18, entitled *"The problem of regulating the Danube"* The edition was printed and bound at a probable cost of about 100,000 schillings. And what happened?

When the experts discovered that the book contained a paper by Schauberger, they arranged for the whole edition to be pulped, reprinted and republished without Schauberger's work, the 100,000 schillings apparently being of no consequence. Are Austria and the Danube States actually swimming in money? None of the journals published or sold officially contained

---

[86]see *The Learned Scientist* and the *Star in the Hailstone*- Ed.
[87]Published by Wirtschafts-zeitungs-Verlags- Gesellschaft m.b.H, Vienna I, Strauchgasse 1. - WZ.

Schauberger's paper. However, he was able to secure several dozen of the original and subsequently destroyed edition, one of which is now in my possession.

[From *TAU* No. 144, p. 27. - Viktor Schauberger]

When the findings of Ehrenberger's investigation were published, I wanted to refute them factually. Mr Fanto, an eminent businessman and then publisher of *Die Wasserwirtschaft*, demanded a sum of Sch25,000 to publish. Due to the official ban on acceptance of any further articles of mine, he asserted that he would endanger his scientific journal.

When I refused to accept his proposition, a hydraulic consultant was called in, according to whose decision a head of department, engineer Rudolf Reich, was supposed to act as mediator. The result of his mediation was a proposal that a certain Dr Lüwy and a Dr Schmal should censor my future articles. A few days later the article I had written was returned to me with the censor's comment that these sharp attacks had no basis in scientific principle and hence the treatise should not be published.

The Federal Chancellor subsequently invited me to have my views on water ennoblement examined by an independent expert. This expert was Prof Dr Mark, who visited me shortly thereafter and studied the so-called water-ennoblement process. At the first trial a petroleum-like substance was produced instead of high-grade springwater, which smelt bad but did not burn.[88] This remarkable abortive experiment, which in the near future will greatly affect the whole world of technology, was the reason that all rela-

---

[88] Here is an interesting extract from one of Schauberger's writings (*Tau* 145, p.18) relating to this phenomenon: "....and thus I succeeded not only in producing petrol-like explosive aqueous substances and the most noble high springwater from dirty water, but also to recreate an invention, which the well-known physicist Gerard Renault had apparently made in his day (1926) and to which he and his assistant fell victim at the Paris Academy, because he failed to recognise the enormous energies latent in water and in the air." An article in *Der Weg* (7.11.1946, Yr. 1, No.48, p.12) elaborates on this event in Paris: The French physicist Gerard Renault had already occupied himself with the problem of obtaining electricity from the air. In his laboratory in Grenoble he worked night and day on his invention and even the scant information that emerged publicly created a sensation in scientific circles. One day a machine stood in his workshop, which was held to be a wonder. Day after day its wheels turned without any observer being able to determine whence the driving force originated. Neither steam, nor combustion gases, nor electrical motive forces could be detected. The machine stood on its base, completely insulated from the ground and ran independently like a perpetuum mobile. For a while Renault enjoyed the general wonderment and then one day he said, "This machine is in fact driven by electricity, but with electricity from the air! I have solved the problem. With its practical application in a few years we shall achieve a paradise on Earth." The French Academy requested Renault to come to Paris so that he could demonstrate his invention there. The inventor agreed and hastening to the capital with his assistant, set up his machine in the experiment theatre. Just before the demonstration began, he wanted to explain how it worked. His assistant naturally was also present. Suddenly there was a tremendous explosion, flames shot out of the doors and windows and once the fire brigade had succeeded in extinguishing the fire, the machine was a heap of wreckage. Renault and his assistant died in the explosion. They took their secret with them to the grave. - Ed.

tions were broken off. So the measurements taken by assistant secretary Ehrenberger could not be disputed, right up to the present day.

It should not be necessary to point out that my views on water and its influence have never changed. Then as now I maintain that, as it gradually warms up, water becomes tired and stale and loses whatever sediment it may carry. In regard to the reduction in flow-velocity associated with a flattening of the gradient through the deposition of sediment, Ehrenberger was unable to detect any such losses in velocity in the Neuberg flume, because the sand-trap built in front of it prevented the entry of any sediment whatsoever. In this flume in which rifling ribs had been installed, the forward surge of the logs, studied by Engineer Karl Heken and Professor Forchheimer, could not be established by Ehrenberger, because very particular mixtures of water are necessary. This I demonstrated to the Assistant Secretary in a double-torsional flow pipe before witnesses, where instead of wood, stones were actually used: these travelled down the central flow-axis without touching the sides.

[From *TAU* No. 144, p. 29. - Viktor Schauberger]

In relation to the so-called 'Theory of Heat' it must be pointed out that very significant differences of opinion exist: I have a completely different view of the various sources of heat compared to the scientific theories advanced by most researchers. Above all I consider the Sun not to be an incandescent saucepan in the order of 6,000°C, because in such a case the oxygen gases evolving on the Earth would have to stream towards this epicentre of combustion. No oxygen could therefore be retained in the Earth and in water, whose content of oxygen plays a decisive role in river regulation. These oxygen gases, instilled into the water through the Sun's energies, are the essential precondition for the growth of water and its quantitative increase. By these means it is possible for water, through increasing its mass and weight, to overcome any diminishing bed-gradient in the valley.

Only through the *growth of water* can the enlargement of channel cross-sections everywhere apparent be explained, wherein it greatly depends how the hydraulic cross-section is formed at various points in relation to depth. The volume of water discharged into the sea, as is commonly known, is a multiple of the sum of the discharge from mountain springs plus the amount of rainfall, wherein it is also to be considered that much water evaporates and infiltrates *en route* and untold quantities of water are consumed in the growth of the vegetation.

Dr Ehrenberger would appear to be totally unaware of the fact that in this water-growth the evolving hydrogen plays a major role. Through the interplay between oxygen and hydrogen, the energies present in stones are released. In this way that substance can first be formed, which we commonly term 'water'. Through the rubbing-together of pebbles, electric sparks are generated,[89] which combine hydrogen and oxygen gases into *new* water. With this the whole chapter on `heat` is closed for, inasmuch as it has a role to play in metabolic processes, it will also be dissipated in the process. As a result the whole water-body remains *fresh, cool* and *full of life* without refrigerators. Through this process of growth, new water-entities are born, which in today's research centres are not *ur*-generated in the 'original' (form-originating) way and hence cannot be identified. All this was overlooked by Ehrenberger. This explains why, in his calculations, he totally and completely failed to take account of the roughly 3m/sec (10ft/sec) movement of the cold volume of water, the constantly changing conditions of absolute and specific weight, the decisive counter-effects of sediment disintegration or, in a word, the *atomic transformation* here taking place.

If stones cannot be pulverised or if water is unable to obtain its provisions for the journey - the latent kinetic energies inhering in sediment - then it is also no wonder that the river's bread is left lying undigested and the channel-bed silts up. Unfortunately, due to statutory requirements concerning patent applications, it is not possible to disclose more detail and so in this case, the only thing I can do is to take the forever-disputed path of demonstrating the existence of these elemental metabolic phenomena through *practical* examples.

There is hardly a single country in Europe from which representatives from the highest official circles have not already visited me to acquaint themselves with my work. The effect was everywhere the same, namely *icy rejection*.

[From *TAU* No. 144. - Werner Zimmermann]

I sent a copy of *TAU* 142 to Dr. Ehrenberger at the beginning of February, as I did to all other persons mentioned therein, with a request for corrections where applicable. On 25th February he wrote the letter which I published on p.24 of *TAU* 144. On 10th March I visited Dr. Ehrenberger at the Research Institute in Vienna and we discussed the matter for almost two hours. Dr. Ehrenberger asked me, *"Will you also be publishing my letter in Tau verbatim?"*. *"Of course"*, I replied. He nodded approvingly. All the same I expressed my concern as to the accuracy of his calculations and indeed the following morning Dr. Ehrenberger telephoned to ask me to delete this section of his letter. I promised that I would do this in *TAU* 144 and also put the amendments in hand.

Mr Schauberger kept the text of the original letter for a long time in order

---

[89] The triboluminescent emission of negative charge. - Ed.

to answer it. These questions were too important that the facts should be allowed to be suppressed. In Switzerland the experts and the authorities tremble before the infallible authority of the Austrian Assistant Secretary and it was absolutely necessary that this glory should be somewhat dimmed, so that we could finally proceed towards a factual and objective clarification. Hence the provision of this information in the service of truth.

Werner Zimmermann.

[From *TAU* 146, p.23 -Dr. Ehrenberger's Missing Section]

"In the calculation of the energy required to heat the water of the Rhine (500m3/s) by 20°C, inaccuracies have crept in. The calculated 57,000,000 hp (TAU 142, p7) will only then result, if the heating of 20°C should take place in one second only. If we now take as an approximation that in order to heat up to this extent, a period from 7am to 5pm (10 hours = 36,000 seconds) is necessary, then the correct quantity of energy amounts to 1,600 hp as compared with the 57,000,000 hp value stated by Schauberger. However, since the assumption of a warming of the water by 20°C is quite improbably high, the actual quantity of energy will be substantially less."

[From *TAU* 148, p26. - Werner Zimmermann]

The interpretation of Ehrenberger's is wrong. The 57,000,000 hp are produced because the 500m^3 of water flow through the Rhine cross-section in one second, so that they must be warmed in one second. Expressed differently: every second 500 new cubic metres of water must be warmed; the volume of water which flows down the Rhine from 7am to 5pm, i.e. which flows through the relevant cross-section, is 36,000 x 500m^3 = 18,000,000m^3.

[From *TAU* 144, Page 31. - Viktor Schauberger's letter to Dr. Ehrenberger - 12 March 1936]

To the Research Institute for Hydraulic Engineering at the Federal Ministry for Agriculture and Forestry, Severingasse 7, Vienna IX/2. For the attention of Assistant Secretary, Dr. Ehrenberger.

Forgive me for not using the customary form of address in this case. Matters have developed in such a way as to prevent me from using the generally accepted polite form of address until the following questions have been thoroughly and satisfactorily answered. On the 10th instant you informed Mr Werner Zimmermann, who visited you at your office at your invitation, that Engineer Richard Prückner came to you one day in order to tell you that the statements I made in *Die Wasserwirtschaft* were supposedly untrue. In your article entitled "Temperature and the Movement of Water"

(*Die Wasserwirtschaft*, No. 9, 1933) you report on page 7 that these statements were the result of an inquiry addressed to Mr Prückner. I would be obliged if you would clear up this minor inconsistency, because in this case it is not immaterial whether someone storms into an office in a fury or is invited there in order to be able to launch into a factually and personally incorrect diatribe affecting my person.

The reasons why I have been unable to respond to this publication up to now are explained in *TAU* 142 and 144. With regard to Prückner's publication, you are to be informed that I had no idea that these data would be incorporated in your exposé and that these were first made available to me on the 10th instant by Mr Werner Zimmermann. In any case, I only heard about your statements by word of mouth.

With respect to Prückner's article, please be advised that I have in my possession Mr Richard Prückner's handwritten data, in which this gentlemen detailed for me the canal profile, the normal volume of water, the length of the Neuwald Pulpmill canal, the type of render, the height of the walls, etc. These data were requested by Professor Dr Forchheimer, who was so interested in this highly interesting information that he went to Neuwaldegg as fast as possible in order to investigate these phenomena. These investigations factually took place, for which Director Prückner's permission was first sought, for without it no stranger was permitted entry to the mill premises.

Prückner at no time has referred to either a positive or a negative temperature gradient, but of his own free will informed me that in this canal extremely mystifying increases in the volume of water occurred, which under certain circumstances increased by up to double the normal flow. I expressly asked Mr Prückner whether I could publish this information and whether, with his permission, I could afford interested parties the possibility of examining these findings more closely. This permission was granted. I made the manuscript as well as the printer's proofs available to Mr Prückner and for years I have heard nothing further.

Professor Forchheimer asked the then manager, Mr Patta, to confirm the data published in *Die Wasserwirtschaft* and in the Gasthof zu Frein, wrote down word for word what Patta had to say about these then mysterious events, which in the meantime have long since been explained. This hiatus is to be thanked for enabling phenomena to be explained which will change the whole basis of science from the ground up and Messrs Prückner and Patta have done mankind a service, whose significance lies far beyond their understanding. This will be elaborated in greater detail at the appropriate time in forthcoming issues of the *TAU* magazine.

The uninitiated will now ask themselves why such trivialities have been endowed with such importance and because of this you, Sir, in particular, felt it necessary to have me branded throughout the world as a liar. On these grounds I am entitled to address various questions to you, Assistant Secretary, whose answers you should consider extremely carefully, because every word will be crucial and under certain circumstances can lead to unpleasant repercussions, whose ramifications you are quite incapable of conceiving.

Questions:

Are you aware that Prof. Dr Forchheimer and Prof. Hauska of the Technical University for Agricultural Science in Vienna provided expert opinions at the behest of the Austrian Department of Forestry, and that Prof. Forchheimer made measurements of the temperature over several kilometres in the Freinbach and the Mürz, the results of which caused these eminent scientists to extend their sojourn in this interesting district for many weeks?

Are you aware that, before a large assembly of university professors in the lecture rooms of the Technical University for Agricultural Science, Prof. Dr Forchheimer was able to demonstrate on the black-board that water temperature plays not only an important, but actually the principal role in the movement of water?

Are you aware that in order to clarify these important questions Prof. Forchheimer took me to various Czechoslovakian universities in order to discuss these questions with his students, who officiate there as recognised hydraulic experts?

Are you aware that Czechoslovakian Prof. Dr Smörcek immediately took me to see Prof. Schaffernack in order to discuss these matters with him?

Are you aware that Prof. Dr Forchheimer urged me to publish these observations in the "Wasserwirtschaft" and that the professor himself saw to it that my articles were accepted for publication?

Are you aware that the river engineering departments of Vienna, Linz, Prägarten and Bregenz, the Chairs for Hydraulic Engineering in Danzig and other places demanded the immediate withdrawal of these articles otherwise they would officially cancel their subscriptions to this scientific journal?

Are you aware that Prof. Forchheimer remarked that he was glad he was 75 years old, because he would find it difficult to have to be re-educated once more?

Are you aware that the world-famous scientist Wilhelm Exner requested me to deposit the findings of my observations in the Academy of Viennese Scientists in view of the possibility that as a layman my authorship rights and priority might one day be contested?

Are you aware that an Assistant Secretary for Water Resources was summoned, who requested Head of Department, Engineer Reich to negotiate with me?

Are you aware that two Austrian Ministers (Rudolf Buchinger and Dr Kienböck) pledged their word with a handshake that I could continue to work freely and undisturbed should I decide to relinquish my life-time job as Wildmeister to the Prince and enter the Austrian government service?

Are you aware that over 100 academics jointly resolved not to permit my presence in government service and to enforce my dismissal?

Are you aware that a falsified transcript of the proceedings of the Kirchdorf District Council was foisted onto the then Member of the Greater German Parliament, Mr Zarboch, as a result of which Buchinger, the then Minister of Agriculture, was forced publicly to give an account of himself in Parliament?

Are you aware that I was then summoned by His Excellency Seipel, to whom I proved that the bona fide transcript of the proceedings proved exactly the opposite to what was contained in the minutes supplied to all members of the government?

Are you aware that Mr Thaler, the former Minister of Agriculture, instituted a high-level disciplinary inquiry and that Departmental Head Dr Kopecky was entrusted with determining whether the statements made by witnesses before the Senate IV in Salzburg were true and that Mr Thaler personally informed me that even as Minister he was unable to force this investigation through?

Are you aware that shortly thereafter I was offered a large bribe at the Ministry, which I refused to accept, firstly because I was supposed to sign a blank document and secondly, because I wanted no money, but demanded permission to present my case, which I was also able to do later when a Viennese industrialist vouched for me?

Are you aware that this man was informed that his business would be seriously jeopardised were he to support me?

Are you aware that this businessman pledged more than Sch. 1,000,000 as a guarantee that I could reduce the usual costs of timber delivery by 90%?

Are you aware that with the encouragement of Assistant Secretary, Engineer Köber, I stated my preparedness to explain the principles of my system of river regulation publicly at the Technical University for Agricultural Science?

Are you aware that this lecture was cancelled at the last minute by the Rector, Dr Olbrich?

Are you aware that this professor publicly declared before witnesses, that this event was the darkest episode of his whole period as rector?

Are you aware that the Austrian Department of Forestry applied for the principal patent in my name, disputing my claim to it after my dismissal and then bought it after I won a patent action lasting a year, because in the courtroom the star witnesses no longer dared to continue to swear by the accuracy of the previous evidence under the glare of the courtroom lights?

Are you aware the then Federal Austrian Forestry Department had to pay Sch. 5,000 per 1,000 logs after I was able to prove that I could transport this timber over a distance of 30 km in a wild, unruly watercourse simply with the aid of temperatures and that the competent authorities were unable to raft even one log 50 metres?

Are you aware that your articles created great difficulties for me in the German Patent Office, because there I was apparently held to be a liar and a swindler?

Are you aware that at that time I was invited by Reichschancellor Adolf Hitler and that on that very occasion Counsel to the Minister Wiluhn produced a document stating that I was only employed as an overseer in the construction of the Steyrling and Neuberg installations, although I possess 42 patents and was the director responsible for the operation these installations, which before and during construction were declared to be follies and which today still deliver double the quota originally guaranteed?

Are you aware that I was invited by His Majesty the King of Bulgaria and that there too similar slanderous material was sent from Vienna?

Are you aware that I was once invited to enter the service of the German Government by the then Prime Minister Braun?

Are you aware that I have entered into negotiations with the widest variety of Foreign Ministers and that on each occasion the negotiations were always broken off at the last minute due to the receipt of untrue information?

Are you aware that Mr Werner Zimmermann has also been warned repeatedly never to have anything more to do with me?

Is it true, Assistant Secretary, that after a visit from Mr Zimmermann, you implored him not to publish a mathematical rebuttal in the *TAU* magazine that you had sent to him in Switzerland?

Mr Werner Zimmermann promised you that he would comply with your wishes. Whatever Mr Zimmermann promises is also sacred to me and for this reason I will also refrain from publishing the mathematically industrious but uninspired piece of work known to me and thus spare you from ridicule before the whole world.

Mr Ehrenberger, on many occasions you have done me great harm. I bear neither you nor Mr Prückner, nor many others who have equal affection for me any resentment, because by your actions you have helped to make the world aware that water is no lifeless substance, but the blood of Mother-Earth, whose innermost nature will never be able to be explained in research laboratories.

You have done all humanity and me a great favour. Had you not freely offered me your help, I would have had to find another Ehrenberger.

Very soon all the patent arrangements will have been settled and then, Mr Secretary, the world will discover all that has happened here.

Your name will go down in history and after many centuries it will be retold how it was successfully proven that water is not H_2O and that the great laws, such as Archimedes' principle, Newton's laws, Mayer's Law of the Conservation of Energy, the law of supposed heliotropism, etc., were great and fateful errors.

Therefore Dr Ehrenberger, please accept my sincerest thanks for your faithful co-operation over many years, which I must now regrettably relinquish, because in the very near future, what so many have not wished to believe will become commonplace, namely that water is the blood of the Earth and is an organism, which must become sick, if the Earth's blood is destroyed through senseless bank-rectification of the surface and subsurface arteries.

Viktor Schauberger - Vienna, 12th March 1936.

The Learned Scientist and the Star in the Hailstone:

A Rare but True Incident

[An article in *Der Wiener Tag*, No. 3381, Sunday, 16th October 1932, page 2, by Viktor Schauberger.]

In 1932-33 Viktor Schauberger wrote a spate of about 11 articles concerning a number of subjects related to his theories on water, which appeared mainly in Viennese newspapers and professional publications, such *Der Wiener Tag, Wiener Neuester Nachrichten, Neue Freie Presse, Architektur und Bautechnik*, etc. Some of these articles may well have been reformulations of the treatises he originally wrote for *Die Wasserwirtschaft*, for which publication had been refused after Professor Philipp Forchheimer's death. The purpose of this entertaining story, one of Viktor's rare excursions into fiction, was two-fold. On the one hand, seeking every available possibility to disseminate his theories, he used it to explain the formation of hail; and on the other to get his own back in part by poking fun at the thoroughly antagonistic attitude of Dr Ehrenberger, who relentlessly sought to thwart Schauberger's every action and to bring his name into disrepute. Ehrenberger was also the individual principally responsible for seeing to it that no further treatises of Viktor Schauberger's were published in *Die Wasserwirtschaft* after Professsor Forchheimer's death. - [Editor]

A fairly long time ago, an absolutely absurd assertion appeared in a Viennese newspaper, namely,

1) that the generally widely held view that the Sun is a molten, incandescent fireball with extremely high temperatures, and

2) that the definition of water by the equally well-known formula, H_2O,

are the *two greatest scientific misconceptions,* the consequences of which will result in not only the economic collapse of all civilised nations, but also the growing unemployment that now affects ever wider circles.

A few days after the article addressing this theme was published, a champion of science appeared in the editorial office of the newspaper. In extremely unflattering phrases, he informed the editor of the enormity of such irresponsible statements, which were highly conducive to creating confusion in the minds of immature young people. The scientific advocate demanded an immediate retraction, stating that the newspaper had either been sold a pup or had been the butt of a bad joke. The editor refused to accede to this demand, fearing that it would endanger the reputation of his newspaper.

Because of this, the editor made an effort to persuade the learned scientist to adopt a slightly more objective view of the article in question. He pointed out that in the main, as history relates, the most important innovations and inventions had been made by outsiders. He also stressed that even if the two assertions by the author of the article were only partially correct, then a change, such as the world had never experienced before, would have to occur in the whole economic situation.

The editor vividly described the consequences that would inevitably ensue once these scientific misconceptions were clarified and recognised for what they were. He further outlined what an end to the appalling unemployment would imply; millions of people would be saved from of starvation and our children, who can only see a dark and dismal future, would now look forward to an improvement in this dreadful, daily worsening situation.

The editor further tried to show that if these two fundamental tenets of science were actually misconceptions, then not only would the entire educational system have to be changed, but many laws and public institutions too. In such an event all our work in such areas as agriculture, forestry, water and electricity resources management, together with a whole range of findings in physics, chemistry, botany, geology, etc., would be seen to be founded on false premises - that they were wrong. He then drew attention to the increasingly apparent mistrust in our children and their disenchantment with our present methods of working, which had already brought us unemployment and them an absolute hopelessness. He concluded with the comment that even he himself, before he was ever aware of the article, had become convinced that this frightful economic collapse was no accident, but could only be the result of serious errors somehow committed by our leading economists.

Taking considerable pains, however, the learned scientist eventually succeeded in causing the journalist, who was not well-versed in these matters, to change his mind. It was finally agreed that in the near future the newspaper would publish an appropriate refutation of these outrageous assertions, to be signed by several scientists. In addition the editor was requested to reject any such articles in the future.

Satisfied with the result, the elderly man now betook himself homewards, already formulating in his mind the rebuff he was going to prepare for the author of these articles, which were as absurd and they were laughable. In his excitement, the learned scholar failed to notice the approach of a large thunderstorm from which a few hailstones had already fallen on the pavement. Suddenly he was struck on the head by a hailstone the size of a pigeon's egg and with a painful lump on his head, the old man quickly sought shelter in a nearby doorway.

After a short time the thunderstorm passed. With an increasingly severe headache, the scholar continued on his homeward way and upon entering his house, lay down to rest. From his resting place he espied two hailstones which had entered his bedroom through the open window and which, warmed by the rays of sunlight falling on the floor, now slowly began to melt, creating a small rivulet of water.

As he looked at these two hailstones, larger than hazelnuts, apart from becoming aware for the first time of the danger he had narrowly escaped, his scientific interest was also aroused as he perceived a brilliant, shining crystalline nucleus in the centre of the hailstones, whose origin, as is well-known, has already occupied a large number of clever minds without success. Reflecting on this difficult problem, he became aware that his agonising headache had become far worse. He suddenly hit upon the obvious idea of using such a hailstone as a poultice to assuage the increasingly acute pain caused by the swelling.

The hailstone had hardly been applied to the wounded spot before the pain began to ease, and under its soothing coldness, whilst still watching the remaining hailstone melt on the floor, the man of learning turned his thoughts once more to the problem of the mysterious star in the hailstone. The star in the hailstone began to shine more and more brilliantly until a bright, cold, brain-penetrating pair of eyes bored into those of the recumbent scholar. Without taking its eyes off him, an icy-grey form slowly stood up on the floor, to the amazement of the old man. What at first had appeared to be a puddle on the floor, was actually the cloak in which the increasingly distinct figure was enveloped and which now began to speak as it slowly approached the eminent savant.

"For decades at school you have taught that the Sun is a molten, fiery ball of over 6,000° Celsius. For decades you have described water, the blood of the Earth, with the ridiculous formula, H_2O. All the while that you have disseminated these misconceptions in several influential journals, many of your students have themselves become teachers. Some have spread your doctrines even further afield, while others have put them to practical use in accordance with the laws of the land. More and more your theories are taking root in the minds of millions of people, thereby greatly increasing your authority and standing, because these theories have become the basis of all scientific professions and government intrumentalities. All your fellow human beings are now at their mercy. These theories have spawned a virus that has caused a dying in Nature that is absolutely devastating and without parallel. Because of your theories, not only was the happiness of your fellow men slowly but surely destroyed, but with them you have also disturbed the whole world, including me in my infinite remoteness and peace. You have forced me to come from the very ends of the Earth to fall upon your addled pate."

Shaking his head, the eminent scientist observed his adversary, who quietly continued to speak.

"Were water merely H_2O as you affirm, then there would be *no plants* on your Earth, no animals and hence *no human beings*. For in reality water is neither H_2O, which in all organisms is a poison, nor is it *contaminated with impurities*, as you have so appetisingly suggested. The substances you describe as impurities are actually the substances (carbones) added to water in conformity with natural law, which, raised from the Earth by the Sun, create all life and therefore even you! What right have you to call these vital, life-giving substances - impurities, when you yourself would categorically deny being a product of impurities!

In the following I will tell you the truth for the sake of the multitude of poor people and especially for the many innocent children, who even now are forced to accept your false theories and their consequences at school. As a result of which they must perish pitifully, because through their work, based as it is on false premises, they will inevitably destroy the very preconditions for their own existence.

I will also tell you the truth so that you will be able to solve the problem yourself - a problem whose solution countless people are seeking even today. They will never find it, because they have all been led astray by your misguided theories. In consequence, you are now charged with the inalienable duty, regardless of any repercussions accruing to you personally, to make it known that everything you have so far taught is fallacious and does not conform to the actual facts.

In Nature all life is built of carbones, which consist of plant residues and which, through the action of the oxygen infiltrating into the depths along with the water, are raised from the Earth under the influence of the Sun's heat in order to promote the formative process mentioned above. This process is to be attributed to the 'breathing' in the so-called living space, which with the help of the Sun and the aid of water, takes place in the following simple fashion, which now perhaps may even become clear to you.

As is well-known, water is also the carrier of heat. When the Sun rises early in the morning, warmth spreads through the living space, causing the boundary between the atmosphere and the stratosphere to be displaced upwards. This shift in the boundary layer only occurs within certain limits, however, which enclose the Earth's surface both from above and below in the form of two neutral strata. The position of the neutral stratum itself varies quite considerably within the boundary zone and is altered slightly due to fluctuations in temperature between night and day, and is displaced to a greater extent by sudden changes in weather and at different times of year. With the movement (advance and retreat) of both neutral strata, the oxygen from above and the suitably preconditioned carbones from below will be brought into the living space. In precisely this way, as simple as it is sublime, the heart-beat of the Earth is factually activated. This breathing process is absolutely essential for, because of it, the formative substances are made to co-mingle with the water in the atmosphere. Under the influence of the alternation of night and day, these substances then build the various forms of vegetation, to which in the very widest sense the Sun also belongs.

In passing through the lower neutral stratum, which is characterised by a temperature of +4° Celsius (this concept, of course, is only to be considered on a strictly relative basis), the strongly carbonated water rising from the depths changes its capacity to dissolve and precipitate matter. Having traversed this stratum and arriving in the vegetation zone of the Earth, the water, now under the influence of atmospheric oxygen, must deposit the carbones carried up from below in the form of sediment. In this process the water eventually becomes deficient in carbones and replaces them with oxygen, thereby becoming heavier, and together with the oxygen, sinks back into the depths after the Sun sets. Now deep in the carbone-sphere and exposed to extremely high temperatures, the oxygen becomes very aggressive and actuates the necessary oxidising processes. Under the influence of the Sun, the products of these processes will then be carried up towards the surface of the Earth by the water, constantly ennobling themselves en route. Similar processes take place in the zone above the upper neutral stratum, but in accordance with the conformities with natural law, they occur in an inverted form.

For certain reasons, too lengthy to explain here, the carbones carried up to great heights by the water-vapour, must constantly be further ennobled (etherealised) until such time as their conformity with natural law reverses as they pass through the upper neutral stratum. Here the hydrogen is left behind and the carbones eventually become carrier-less. Having now become carrier-less, the highly etherealised carbones stream towards the Sun, where due to the effect of extremely low temperatures the most refined and ennobled oxygens are concentrated. Here the high-grade carbones will be transmuted into their most highly transformed state - *into pure energy*. The rays emitted from this epicentral focus of oxidation undergo a warm oxidation on their return journey, thus participating in the opposite process that occurs on their way towards the Sun. The region from the Sun to the neutral stratum, which is now under the influence of a negative temperature gradient, encompasses that part of the radiation spectrum in which the rays from the Sun are not immediately perceptible as well as the bandwidth where these have in part already been transformed into light rays. The region from the upper neutral stratum down to the Earth's surface, which equally exhibits a negative temperature gradient, is to be described as the region where conversion into heat takes place.

What happened to the hydrogen, which was gradually left behind en route towards the Sun? Due to the extremely low temperatures at this high altitude, it crystallised into fine ice-particles, having attained its highest state of aggregation after surpassing the upper neutral stratum. A process now takes place, which is of extreme importance. Each one of these minute particles of ice carries within it a nucleus of *very dense oxygen*. Due to the absorption of this highly condensed and crystallised oxygen, the fine-ice crystal increases in weight, overcoming the considerable buoyancy active at these high altitudes and the fine-ice body begins to fall according to the law of gravity. When this fine-ice crystal reaches the lower, warmer zones it thaws. Upon passing through the upper neutral stratum on its way down, the carrier - hydrogen - alters its state of aggregation and dissolves. Becoming free in the process, the oxygen now mixes with the air, simultaneously reducing its aggressivity and thus forming a less complex form of oxygen.

An analogous process, even if in inverted compliance with natural law, we also find in the Earth. Accompanying the atmospheric water, not only do the denser forms of oxygen succeed in infiltrating the Earth, but to a certain extent the very valuable carbones do as well. Without the interaction between these substances, the solution of the salts present in the root-zone and lower down cannot take place correctly. With the aid of these aggressive carbones and oxygenes, the salts in the ground will now be dissolved and transformed into various forms of nitrogen. These substances alone can be properly processed by the plants. Taking place with the aid of sunlight, this transformation enables the fluids in the water rising in the vascular bundles be transformed into sugars, starches or cellulose. Some of the refined and ennobled carbones, however, depart in the direction of the Sun, as described earlier. Without these natural processes, the build-up of any vegetation would be impossible. When these great conformities are disturbed, then a degeneration takes place, which is synonymous with the qualitative decline of all forms of vegetation.

With a sudden drop in temperature, also the result of disturbances, sometimes the previously described fine-ice crystals do not have time to melt and can actually reach the Earth. Due to the extreme cold they emanate, they gradually become sheathed with a solid layer of ice on their way down, which even the air's resistance cannot remove. In this manner larger ice-bolides are formed, each of which carries the mysterious star in its centre. One of them wounded you today!

As you may recall, Hörbiger, that brilliant researcher, also spoke of the enormous degree of coldness possessed by these granules of fine-ice, which come from the coldest and highest regions of the stratosphere. In any case, even without Hörbiger it ought to have been obvious to you that the well-known gas nebulae can only be formed at low temperatures in accordance with the simplest laws of mechanics, and hence in these zones the high temperatures you always talk about cannot exist. But all you scientists are a very peculiar breed! Every one of you consistently places himself beyond all connection with the world around him and because of your one-sided specialisation, you increasingly divorce yourselves from all reality and the actual workings of Nature. Inevitably, therefore, each of these areas of specialisation has produced certain phoney findings, whose application and effect, however, have changed the face of the Earth. In their totality, these effects have become so large that they even affect the natural development of children, forcing their thinking and reasoning, even their whole being into unnatural ways. In this process these young, developing human beings are ultimately robbed of the inner connection with Nature they received at birth. Only thus could this dreadful curse of decay have engulfed a humanity that has set the force of one-sided *mechanistic* technology against a second stronger, and far more efficient natural force, the *organic ecotechnology* in Mother-Earth. The only possible outcome of your monstrous aberrations is the enormous misery of vast multitudes, which will greet you at every street corner in the most pitiable fashion from the eyes of children, mothers and old men."

This is where the being's explanations ended. The factual explanations of the learned scientist's adversary, especially the line of argument using the laws concerning the dynamics of gases, had irritated him enormously. Seeing himself slain with his own weapons, his opponent's final rebukes completely robbed him of his composure. Furious, he sprang up, only to sink back immediately with a cry of pain - and woke up. At the point where just a quarter of an hour earlier he had placed the improvised ice-pack, he now felt a burning pain and quickly summoned the doctor.

The doctor soon diagnosed a severe cold in the head and applied his remedies accordingly. Despite the most careful attention, the condition of the patient deteriorated and the fever became more and more intense. The doctor was bewildered by this inexplicable phenomenon. Upon the doctor's repeated questioning, the sick old man related his dream. Slowly it became clear to the doctor that his treatment had been inappropriate and that he was not here concerned with a chill, but rather with an inflammation of the tissue.

While the less complex atmospheric oxygen is able to bring about transformations of far-reaching importance when it comes in contact with the rising carbones in the Earth, the oxygen contained in atmospheric water causes even more far-reaching transformations. What dangers, therefore, must the high-grade, aggressive oxygen contained in fine-ice crystals, which already approximates rays of energy, trigger off in the over-heated blood of an earth-dweller with a headache!

Impressed by this new understanding, the doctor at once became aware of the enormous danger that now threatened the life of his learned patient. In direct consequence of the release of oxygen from the ice-crystal onto the feverish head of the eminent scientist lay the danger of the development of cancerous decay, a phenomenon we are unfortunately already able to observe throughout our forests. The spread of cancerous diseases in trees is the result of carbone-deficient and oxygen-enriched water rising in the trees' capillaries due to the exposure of the most shade-demanding species of timbers to the light. Having become aggressive due to the warming effect of direct solar radiation, the oxygen now promotes the development of swellings through the decomposition of the tissues. This ultimately causes the cancerous tumours that have now spread through almost our entire high forest.

Because of the doctor's new perceptions, his method of treatment changed and the eminent scholar recovered. Rumour has it that for months and months he has been working on the rebuttal, which at the time he had so unwisely promised the editor.

Viktor Schauberger, Vienna 1932.

Appendix

PATENT APPLICATIONS

It was Viktor Schauberger's customary practice to seek immediate patent cover for all his inventions and devices, of which the following principally concern those related to water. Unfortunately, just about all of the patents he applied for during World War II are no longer available and therefore, while his various devices are described in a number of his writings, no visual representation of them exists. This makes their proper description an extremely difficult task. - [Editor]

Specification Of Patent No. 134543

AUSTRIAN PATENT OFFICE
SPECIFICATION OF PATENT No. 134543
Class 47f. Issued 25th August 1935.

VIKTOR SCHAUBERGER IN VIENNA
THE CONDUCTION OF WATER IN PIPES AND CHANNELS.
Application date: 12th August 1931 - Patent applies from: 15th April 1933

The object of the invention is a system of water conduction, which in contrast to smooth-walled conduits, channels, pipelines and the like, promotes an increase in the transported volume of water. In the opinion of the inventor, which forms the basis of his invention, turbulent phenomena in conventional systems of water conduction are in part caused by differences in the temperature of the various water-strata, principally because the velocities of the water-masses flowing along the pipe-walls are substantially different from those of the more central strata, causing vortical phenomena at their mutual interface.

In order to inhibit sedimentation, it is claimed that projecting, turbine-blade shaped elements (guide-vanes) should be incorporated, which are inclined from the walls towards the centre. Each of these should be so curved as to direct the flow of water from the periphery towards the middle. It is also to be noted that the inner walls of the pipe are to be provided with raised and curved, rib-like projections in order to impart a rotational motion to the water.

The present invention concerns a further development of these measures with regard to the aims mentioned at the beginning. In the attached diagram, various aspects of the invention are depicted. *Figure 1* shows an isometric view into the pipe, *Figure 2* an oblique view of a single guide-vane, viewed in the opposite direction to the current and *Figure 3* the same is viewed at right angles to the direction of flow. *Figure 4* depicts how the invention is to be installed in a channel. *Figure 5* shows a cross-section of a guide-vane incorporating rifle-like fluting aligned to the direction of flow.

In pipe *1*, a series of guide-vanes *2, 2', 2"* are placed along the curved lines of multiple helical paths *3, 3', 3"*. The latter are shown in broken lines. The guide-vanes themselves are curved in the manner of ploughshares and project from the walls of the pipe in such a way as to deflect the water towards the centre of the pipe, at the same imparting a rotational motion about the pipe axis.

In *Figures 2* and *3*, which give oblique

and side views of a guide-vane, the straight, dotted arrow indicates the direction of flow in a smooth-walled pipe, whereas the curved, solid arrow shows the path of the water filaments deflected by the guide-vane. Similar guide-vanes can also be installed in channels. In this case the guide-vanes are not placed along a helical path, but one directly behind the other and as shown in *Figure 4*, are arranged symmetrically on both sides at equal heights and directly opposite each other.

The vane in *Figure 5* is provided with rifle-like fluting on its guiding surface, through which in the course of such spiral motion, the forward movement of the water will also be given a vertical lift. Pipes incorporating this type of guide-vane are especially suited to the transport of matter heavier than water, such as ores and the like.

PATENT CLAIMS

1. The conduction of water in pipes and channels is characterised by the proposed incorporation of turbine-blade-like elements (guide-vanes), projecting inwardly from the surface of the pipe and/or channel walls towards the centre of the same. Each of these elements is so curved as to direct the water from the periphery towards the middle of the conduit, such that in pipes, the guide-vanes are mounted along multiple spiral paths, whereas in channels, these are placed one directly behind the other and arranged symmetrically, both opposite each other and at equal heights on each side of the channel.

2. In accordance with Claim 1, the conduction of water in pipes and channels is further characterised by the proposed incorporation of rifled fluting on the guiding surfaces of the vanes, which runs parallel to the direction of flow and which directs the flow from the periphery of the pipe towards the centre.

See Fig. 8
Note: Figures referred to in patent text relate to those indicated in fig. 8 of this book.

Specification Of Patent No. 138296

AUSTRIAN PATENT OFFICE
SPECIFICATION OF PATENT
No. 138296
Class 47f. Issued 10th July 1934.

VIKTOR SCHAUBERGER IN VIENNA
THE CONDUCTION OF WATER.
Supplementary Patent to Patent No. 134543
Application date: 2nd November 1932 -
Patent applies from: 15th March 1934
Longest possible duration: 14th April 1951

The present invention concerns a further development of the system of water conduction described in Patent No. 134543, in which turbine-blade shaped elements (guide-vanes) project inwardly from the pipe walls towards the centre of the pipe and which are so curved as to direct the water from the periphery towards the centre, wherein, according to the original patent, the essential aspect of the invention consists in the positioning of guide-vanes along multiple helical paths.

In accordance with Patent No. 134543, the particular form of the guiding surfaces of the vanes is such that they are provided with rifled fluting, which follows the direction of the current. This invention concerns a further development of these guide-vanes, whose purpose is to enhance the fast forward movement of the central core of water in relation to the flow in the peripheral zones.

The normal restrictions to the flow in the peripheral zones leads to turbulent phenomena in the boundary layer between peripheral and core zones and influences the proper formation of the core zone unfavourably. The purpose of the present invention is to divide the peripheral zone into separate, individual vortical formations, which due to their inner stability, in a manner of speaking, become stable structures with only a slight tendency to disintegrate. In their aggregate these provide an outer envelope of water, which enhances the forward acceleration of the core-water.

These vortex-creating elements are twisted like wood-shavings, so that two direction-controlling surfaces can be created, in essence according to **Figure 1**. The purpose of these two surface-elements is to impart a torsional motion to the water filaments in the zone of peripheral flow, the direction of which is indicated by the arrow 3, so that a subordinate spiral motion is created within the general spiral motion of the whole water body. In *Figure 1,* the top view of the invention is shown. *Figure 2* shows a perspective of the invention, viewed in the opposite direction to the flow of the current. *Figure 3* shows the shape of the invention when flattened out.

The guide-vane 2 is arranged in pipe 1 along multiple helical paths in accordance with Patent No. 143543. When departing from portion 5 of the guide-vane, in each case the water filaments are always imparted a movement directed towards the centre of the pipe cross-section. The flow of water will be enhanced by the ribs 6 and because the ribs converge conically, the water becomes compressed, which should likewise impel the fast-moving transported matter towards the centre. The guide surfaces can also be assembled from separate elements.

PATENT CLAIMS

1. In accordance with Patent No. 134543, the conduction of the water is characterised by guide-vanes projecting inwardly from the wall-surfaces of the pipe towards the centre, such that, akin to wood-shavings, these turbine-blade shaped elements are twisted so as to create two co-acting fin-like surfaces. The first of these surface-elements (upstream element) separates the peripheral zone of the current from the core zone and the second element (downstream element) additionally imparts a convoluting motion to the separated bundle of water filaments due to the twisted shape of the guide-vane surfaces, whereby the peripheral zone will be divided into individual, stable, vortical structures.

2. In accordance with Claim 1, the conduction of the water is characterised by the fact that, when flattened out, the ribbed guide-vanes possess an almost rhomboidal form *(Figure 2),* whose diagonally opposed obtuse-angled corners are bent over towards the opposite corner *(Figure 3).*

See Figure 9

Note: Figures referred to in patent text relate to those indicated in fig. 9 of this book.

Specification Of Patent No. 142032

AUSTRIAN PATENT OFFICE
SPECIFICATION OF PATENT
No. 142032
Class 85c.   Issued 11th June 1935.

VIKTOR SCHAUBERGER AT HADERS-DORF-WEIDLINGAU (LOWER AUSTRIA) PROCESS FOR THE PRODUCTION OF SPRINGWATER-QUALITY DRINKING WATER.

Application date: 22nd February 1934 -
Patent applies from: 15th January 1935

It is known that mineral water is produced in such a way that salts are mixed with any variety of hygienically impeccable tap or spring water and that gases are introduced under a pressure of at least 2-3 atmospheres, usually under a higher pressure, however. It is also known that to produce sodawater, carbon-dioxide is introduced into the water under a mechanical pressure of about 12 atmospheres, creating a corresponding concentration of so-called free carbonic acid, which appears to be bound in the water by mechanical means only. A similar process is involved in the manufacture of "fizzy pop".

In the artificial manufacture of mineral water, carbon-dioxide is similarly introduced under greater or lesser pressure, which in any event, however, is greater than 1 atmosphere, wherein a sufficient quantity of certain salts is admixed as the taste of the mineral water demands. A further well-known method of producing effervescent drinks consists in dissolving any easily soluble carbonate (sodium bicarbonate for example) and adding weak acids (such as tartaric or citric acid), through which free carbonic acid is similarly evolved, which is the cause of the prickly taste of the drink thus created.

The present case, however, concerns the production of water of a kind in which carbon-dioxide is not merely concentrated in unbound form, but is contained in bound form, and in every respect is similar to good, high-grade springwater, wherein the manufacturing process emulates the processes in Nature as far as possible.

Sterilised water flows down pipe *m* under a cold, mercury-vapour lamp and mixes with the salt solution coming from duct *l*. In vessel *C* the salts are actually dissolved and well stirred by mixer *g*. The quantity and type of salt is naturally dependent on the composition of the parent water, which in most cases will be any variety of surface water exhibiting certain permanent degrees of hardness. On the other hand, through the addition of salts, the hardness of the water to be produced should not be caused to exceed 12 degrees, otherwise the product would be difficult for industry to use. For every 10 litres of any average quality parent water, 1 litre of salt solution is prepared in which about 0.02g sodium-chloride, 0.02g magnesium-sulphate, 0.02g sodium-biphosphate, 0.008g potassium-nitrate and 0.2g calcium-oxide are dissolved. The type and quantity of these salts are the result of several hundred experiments. Since in the beginning the calcium-oxide is only partially dissolved in the water, and on the other hand, the evolving calcium hydrate is sensitive to atmospheric carbon-dioxide, the vessel is sealed against light and air. In order to maintain a constant rate of discharge from vessel *C*, it is placed under a constant pressure of 0.1 atmospheres = 1 water-column metre. The concentrated salt solution is continuously added to the water in pipe m drop by drop and the mixture of both fluids flows into the atomising unit *D*, where it is sprayed into the interior of the vessel through the small holes *n* in the pipe, whereas the already previously prepared carbonated water is sprayed outwards from atomising pipe *k*.

The droplets of both types of water then precipitate, mixing together drop by drop en route in the same way that in Nature every drop on its way down into the Earth first dissolves salts and absorbs gases. This quantity of water now flows into unit *E* containing tulip-like glass bulbs and always rises up in the outer glass tulip, subsequently to be made to descend in the inner glass tulip, in order to rise further through the innermost riser pipe into the next higher outer tulip. During this process the water therefore follows a meander-like path en route to the processes described later. The gas, that is to say, predominantly the carbon-dioxide, always accumulates in the upper part of the tulip and once the pressure has risen sufficiently, will then be continuously injected into the path of the rising water in the innermost riser pipe via the tube *r*, in which the very finest jets have been incorporated, so that any carbon-dioxide not previously bound, will be forced to bind itself with the water. On the axis of this component of the apparatus, an alternating series of gold and silver laminates are attached, but which are insulated from each other. A certain potential is created between both metals, giving rise to the weak ionisation of the water.

Further along its path, the water enters the main mixer *F*. This consists of an externally thermally-insulated cylindrical metal vessel, silver-plated on the inside, incorporating an Archimedes screw-like auger, which rotates in the opposite direction to that in which the spiral is generated. In addition, cooling coils are affixed to the surface of the spiral blades, which cool the water to +4°C, whose ennobling process begins at 17°C. The effect of this drop in temperature is of integral significance to the actual process of ennoblement. By being cooled, on the one hand the water's capacity to absorb gases will be increased and on the other, the substantial binding of free carbon-dioxide (without application of pressure) in such ample measure is only possible through cooling.

The $Ca(HCO_3)_2$ (calcium-bicarbonate) is an extremely labile compound, which is responsible for the concentration of the so-called dissolved carbon-dioxide (bound carbonic acid) in the water. However, the substantial binding of $Ca(HCO_3)_2$ and with it the proper and effective binding of the carbon-dioxide (carbonic acid) to the water is only possible with the right degree of simultaneous cooling. In this connection, the starting temperature of the water should not exceed about 20°C and the end temperature must approximate +4°C. Attention should also be paid to the rate of cooling and if this takes place too rapidly, sufficient binding will likewise not be achieved. Before leaving this vessel, the water must again pass by gold and silver laminates, whose function has previously been described, finally reaching the storage vessel *I*, which is divided into two chambers *G & H*. The water only reaches chamber *H* by overflowing from chamber *G*, for the following reasons. With the treatment of the water in the way outlined above, certain delayed reactions occur. Only when these have ceased can the water be described as completely ready to drink. It is also necessary that this process should take place entirely in the dark (insulated from light), since experiments have shown that similar processes of ennoblement under the influence of light yield a more inferior water.

PATENT CLAIMS

1. The process for the production of springwater-like drinking water is characterised by the mixing of sterile water, alloyed with small quantities of various salts in a finely atomised state, with equally finely atomised carbonated water, in which the mixed product is cooled on an extended route comprised of cross-sectional profiles of various shapes and forms.

2. In accordance with Claim 1, the atomisation of the salt-alloyed water is effected by means of a perforated systems of tubes, through which it is expressed into a sealed, dark, air-tight vessel and as it falls towards the bottom of the vessel, mixes drop by drop with similarly atomised carbonated water, also issuing from perforated tubes.

3. In relation to Claims 1 and 2, the mixed product flows through an apparatus isolated from light and air, in which the water describes a meander-like course and is alternately conducted through narrow and broad cross-sections, whereby the carbon-dioxide precipitated at the broader cross-section is re-injected into the water at the narrower section.

4. In relation to Claims 1 to 3 of the process, the water is further conducted through a mixing unit in which it must describe a screw-like path, wherein the screw turns in the opposite direction to the turn of the thread and on whose surfaces cooling coils are attached for the purpose of raising the water's capacity to absorb gas as it is cooled towards the anomaly point of +4°C.

5. In order to carry out the process, certain equipment is necessary according to claims 3 & 4, wherein it is hereby stated, that an alternating series of gold and silver laminates are affixed at well-insulated positions.

See Figure 3
Note: Figures referred to in patent text relate to those indicated in fig. 3 of this book.

Specification Of Patent No. 117749

AUSTRIAN PATENT OFFICE
SPECIFICATION OF PATENT
No. 117749
Class 88.   Issued 10th May 1930.

VIKTOR SCHAUBERGER IN PURKERSDORF, LOWER AUSTRIA.
A JET TURBINE.
Application date: 21st December 1926 -
Patent applies from: 15th January 1930

The object of the invention is a hydro-electric device, which exploits the kinetic energy of a water jet for the purposes of generating electricity.

The invention is characterised by a cone-shaped rotor, whose apex points towards the outlet opening, and rotates about an axis common to both rotor and water jet. The outer face of the cone is formed of upward-facing, concave, corkscrew-like blades. In this way the water-jet is split up and deflected from its path and imparts its full force to the rotor, so that, with the appropriate proportions between the height of the cone and the width of its base, and a suitable pitch of the blades, the size of which is dependent on the velocity of the impacting water-jet, the water flows from the machine quietly without creating spray.

An example of the arrangement of the invention is schematically depicted in the diagram.

The rotor, whose axle *1* is parallel and common to the axis of the jet exiting from the jet-pipe *2*, is formed of corkscrew-like blades *3*. The ends *4* of the blades *3* are curved upwards slightly towards the impacting water-jet so as to deflect the jet and to effect the greatest possible transfer of its kinetic energy to the rotor.

In the jet-pipe *2* screw-like ribs *5* are incorporated, which, according to observations, increase the velocity of the exiting water-jet and the efficiency of the device.

PATENT CLAIMS

1. The jet-turbine is characterised by a cone-shaped rotor positioned in the axis of the water-jet, by means of which the water-jet is split up. Corkscrew-like blades *(5)* are incorporated around the cone's periphery *(7)*.

2. In accordance with Claim 1, the jet-turbine is further characterised by a jet-pipe *(2)* incorporating rifling ribs *(5)*, which impart a spin to the rotor in the direction of its rotation.

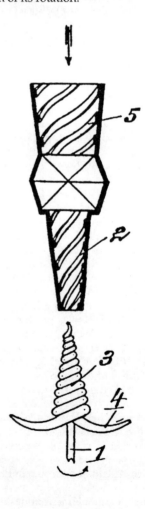

Specification Of Patent No. 113487

AUSTRIAN PATENT OFFICE
SPECIFICATION OF PATENT
No. 113487
Class 84.   Issued 10th June 1929.

VIKTOR SCHAUBERGER IN PURKERSDORF, LOWER AUSTRIA. A DEVICE FOR TORRENT CONFINEMENT AND RIVER REGULATION.

Application date: 31st January 1927 - Patent applies from: 15th January 1929

The object of the invention is a device for the purposes of torrent confinement and river regulation, by means of which the velocity of the water can be braked in such a way that the transported sediment can engender no hazardous, destructive effects and the movement of the water can be so influenced as to displace the theoretical flow-axis towards the middle of the channel.

The attached drawing depicts the object of the invention schematically and *Figure 1* shows the installation of such a braking, flow-guiding device in the form of brake-groins installed at right angles to the direction of flow.

The brake-groins *1* are desirably made out of reinforced concrete and are anchored into the ground by the downwardly projecting stumps *2* shown in *Figure 1*, to prevent their being dislodged by the onflowing water. In an upstream direction these brake-groins incorporate a concave, fluted wedge-shape (*Figure 5*), onto which the water flows and by means of which it is lifted and directed towards the centre of the channel, thus dissipating a great deal of its momentum and rendering it incapable of transporting larger rocks or stones.

These brake-groins are installed at greater or lesser intervals in the stream-bed, according to the steepness of the gradient. In order to displace the theoretical flow-axis towards the centre of the channel during the course

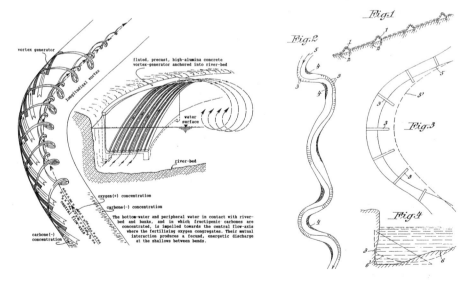

Figs. 33 & 34: Note: Figures referred to in patent text relate to those indicated in fig. 33 of this book. Fig. 34 is the translator's interpretation of the above patent description, the latter seeming to be at variance with the patent diagram (fig. 33). The design shown in fig. 34 is more in accord with the curve generators shown in figs. 4, 5 & 6.

of flow and corresponding to the purpose of the invention, these water-braking devices are installed on the sides of the channel at right-angles to the direction of flow in those locations where pot-holing and the undermining of the riverbanks occurs or is likely to occur, as shown in *Figures 2* and *6*. In *Figure 2* the brake-groins are indicated by the number 3, whereas the deposition of sediment occurring on the opposite side of the channel is indicated by the number 4. The flow-axis desirably to be displaced by these installations is shown by the arrowed line indicated by the number 5.

Figure 3 depicts the device at a larger scale and *Figure 4* shows the cross-section through the same. The essential shape of the device is triangular (*Figures 4* and *5*) and its active surface rises towards the riverbank and gradually projects towards the centre of the channel. The function of these devices is particularly apparent in *Figure 4* in which the solid line 6 shows the bed-profile prior to the installation of the device and the dotted line 6' indicates the profile ultimately produced.

Between these braking-groins 3 the transported sediment is deposited, creating a zone of dead-water near the bank, which serves as a buffer and keeps the flowing water-body away from the bank, thus preventing the bank from being undermined (*Figure 3*). In *Figure 3* the solid line 5 shows the flow-axis before installation of the devices and the dotted line 5' shows the displaced flow-axis due to the action of the invention.

PATENT CLAIMS

1. The device for torrent confinement and river regulation is characterised by the concave fluting on the upstream side so that the on-flowing water is deflected upwards and backwards, or towards the middle of the channel.

2. In accordance with Claim 1 the device is further characterised by its triangular shape, which projects from the bank at right-angles to the current flow.

Description Of Patent No. 136214

AUSTRIAN PATENT OFFICE
DESCRIPTION OF PATENT No. 136214
Class 84.   Issued 10th January 1934.

VIKTOR SCHAUBERGER IN PURKERSDORF NEAR VIENNA CONSTRUCTION AND EQUIPMENT FOR REGULATING THE DISCHARGE FROM DAMS AND FOR INCREASING THE STABILITY OF DAM-WALLS

Application date: 23rd April 1930 - Patent applies from: 15th August 1933

The invention concerns the design of plant and associated equipment for regulating the downstream channel of reservoirs and for increasing the structural stability of their barrage-walls. In particular, the invention consists in the fact that a mixture of heavy- and light-water, which is suited to and dependent upon the external temperature, can be conducted from the reservoir into the drainage channel automatically and in such away that, as circumstances demand, the heavy-water to be discharged into the drainage channel can be diverted to cool the valley-side of the barrage-wall by over-trickling it with heavy-water.

It has become evident that in all hydraulic practices applied to the drainage of water in channels, an important factor has been disregarded, namely the temperature of the water in relation to ground- and air-temperatures, as well as the differences in temperature in the flowing water itself. Furthemore it has also been determined that the existing and constantly changing differences in temperature influence the movement of the water decisively. Inasmuch as the natural channel is subdivided by artificial constructions, such as dams, weirs and the like, and the discharge therefrom is either via bottom-sluices (which discharge heavy-water with a temperature of about +4°C only) or via the spillway (whereby the downstream channel is supplied with the currently highest temperature water), disturbances develop in the downstream channel, which in particular give rise to curves in the channel and to the destruction of the riverbank. However, if water of a temperature corresponding to the ambient external temperature, i.e. correctly tempered water, is discharged into a given channel, then as circumstances dictate, the water-masses can either be braked and their sweeping-force reduced or conversely, they can be accelerated and their sweeping-force increased. Instead of regulating the channel with bank-protecting structures, whose effect is only local, it is therefore possible to bring about the disturbance-free drainage of the water-masses solely through the regulation of the right water-temperatures; that is, through the automatic establishment of an enduring state of equilibrium in the water itself. Widening of the channel through the deposition of sediment, or the ejection of the same (gravel banks), and fissures in the riverbank, especially at the bends, can be prevented by properly designed and equipped dams, and incorrect drainage conditions corrected. Through the appropriate adjustment of the mechanisms incorporated in these dams for controlling the discharge of light- or heavy-water, the temperature gradient corresponding to the ambient external temperature can be re-established and in this way the danger of flooding in particular can be almost completely averted.

Concurrently with the regulation of the drainage channel, the stability of the structure required for this purpose, namely the specially designed barrage-wall of the reservoir, can also be increased in a manner whereby the pores in the wall-structure are sealed through the cooling of the water-particles infiltrating into the wall from the reservoir, thereby removing the cause of the wall's destruction. With a reduction in temperature, the light-water infiltrating into the wall-pores loses its ability to transport and dissolve salts and other substances, until at a temperature of +4°C it reaches the condition where its dissolving power is at minimum and the filter-action of the wall is greatest. Through the cooling of the valley-side of the barrage-wall by overtrickling it with +4°C heavy-water, the light-water infiltrating from the reservoir is cooled and precipitates its dissolved substances into the pores, thereby sealing them. The water-tight sealing of the wall-pores is achieved within a few weeks, thus making any further safety precautions against the destruction of the wall superfluous. Should the aforementioned cooling of the valley-side of the wall be omitted, then the light-water infiltrating into the wall from the reservoir will be warmed from the valley-side of the wall, in particular by solar irradiation, thereby gaining in dissolving power vis-à-vis the solid particles of the construction material. The pores will be leached out. With increasing enlargement of the pores, the explosive action of frost will also be greater. Fissures will develop in the wall, which permit the entry of more water not only as a result of hydrostatic pressure, but also due to current-pressure, until such time as the structure of the wall, particularly at the height of the normal water-level, is completely destroyed.

The diagram depicts an example of the design of the installation, namely the barrage-wall of a dam. *Fig. 1* shows a cross-section and *Fig. 2* the plan, whereas *Fig. 3* is a detail showing the discharge control-mechanism in section.

For the purposes of regulating the amounts of cold heavy-water and warm light-water, sluices O in the barrage-wall K of the reservoir B are incorporated on both sides of the same, whose sluice-gates T are operated by a temperature-controlled floating body G. The rising pipes W connect the sluices O to the main spillway K_1 of the barrage-wall. Diverter-pipes U^1, U^2 and U^3 are located at various heights, which branch off from the rising pipes W and are controlled as required by stop-valves V^1 and V^2. These diverter-pipes lead to the valley-side of the barrage-wall K and discharge into their respective, horizontal troughs. At the base of the valley-side of the barrage-wall K, an upwardly curved structure K^3 is incorporated for the purposes of creating vortices and for the better mixture of the water flowing over the wall.

The blades of the sluice-gates T rest on a sill recessed into the bottom of sluices O and their water-tight closure is effected by pressure-alleviating rollers set in vertical grooves. By means of a connecting-rod F, situated in a shaft in the side-wall H, the sluice-gate T is attached to the floating body G, which can be shaped like a diving-bell for example. In the side-wall H at various heights above sluice O, pipe-shaped openings A are incorporated, which communicate between the shaft in which diving-bell G floats and the open water of the reservoir. When sluice-gate T is opened, communication is also achieved between riser-pipe W and the reservoir through the filling of the riser-pipe, which relieves sluice-gate T from one-sided pressure, thus ensuring its most friction-free operation. Being constructed preferably of timber, sluice-gate T can therefore be precisely adjusted to the carrying capacity of diving-bell G, so that its free movement under all water conditions is assured. Diving-bell G, whose position on connecting-rod F is variable, can therefore be set to float at any desired height. In the lid of diving-bell G there is a closable air-vent P, through which, if opened, compressed air within the diving-bell can escape, causing sluice-gate T to shut immediately. By means of a vertically calibrated pipe R, open at both ends, the water-level inside the diving-bell can be set to any desired height depending on the depth at which the bottom of pipe R is fixed. When diving-bell G is completely submerged with no internal air-cushion, it can be raised through the supply of compressed air via pipe R by shutting air-vent P, thereby enabling sluice-gate T to be raised. During normal operation the air-cushion enclosed within the diving-bell is in close contact with the atmosphere via the diving-bell wall, so that, particularly in the case of metal walls, the external temperature will exert an influence on the volume of the air-cushion. Depending on the external-temperature-related increase or decrease in the volume of the air-cushion in diving-bell G, sluice-gate T will either be raised or lowered. The amount of heavy-water conducted to the valley-side of the dam-wall via sluices O, rising-pipes W and diverter pipes U^1, U^2 and U^3, and discharged into the channel, will therefore vary according to the external temperature. The light-waterflows over a special spill-way-structure M above the top of the dam-wall and down into the channel.

The thorough mixture of heavy- and light-water will not only be facilitated through the upwardly curving structure K^3 at the base of the valley-side of the barrage-wall, but also through the conduction of heavy-water via the horizontal diverter-pipes U^1, U^2 and U^3 and their respective troughs into the path of the vertically falling light-water; their intimate mixture being achieved by means of the vortices created artificially in this way. As each individual diving-bell G is irradiated by the sun, the respective sluice-gate T will be further raised and in this way a greater percentage of heavy-water will be added to the light-water flowing over the top of the barrage-wall at M, whereas with cool external temperatures, sluice-gates T will be nearly or completely closed, allowing only warm light-water to overflow into the channel.

The heavy-water conducted to the top of the dam wall at spillway K^1 for purposes of better mixture can simultaneously be employed to increase the stability of the barrage-wall. Once construction of barrage-wall K has been completed, the lower, valley-side portion of the barrage-wall K will be over-trickled with heavy-water exclusively by means of diverter-pipe U^2 for instance, for which purpose diving-bell G will be so adjusted that the sluice-gates T will remain open constantly. At this juncture an overflow over the top of the dam is not appropriate and the heavy-water supply will be conducted directly to the channel via sluice O. The heavy-water overtrickling the valley-side of the barrage-wall now cools the wall from the outside to such an extent that the light-water percolating into the wall-pores from the reservoir deposits its dissolved substances and seals them off. After the lower portion of the barrage-wall has been sealed, the heavy-water can be conducted to the upper portion of the barrage-wall via diverter-pipes $U3$, which can then be sealed in a similar fashion. This sealing process, during which the wall-pores become water-tight, may require several weeks, depending on the quality of the construction material. Once completed, no further dangers are to be feared, even during normal operations. After the wall has been sealed, the special spillway-structure M, which need only be made of steel and placed atop the wall temporarily, can also be removed so that the light-water, instead of discharging over spillway M, will overtrickle the top of the dam-wall K^1, thereby preserving and protecting the wall-structure on the valley-side.

PATENT CLAIMS

1. The design of the installation for regulating the downstream channel of reservoirs and for increasing the stability of their barrage-walls is characterised by the provision of equipment by means of which a mixture of heavy- and light-water, suited to and dependent upon the external temperature, is automatically discharged into the downstream channel.

2 In accordance with Claim 1, the design of the installation is characterised by the incorporation of mechanisms whereby the valley-side of the barrage-wall K of the reservoir can be cooled by overtrickling it with heavy-water.

3. In accordance with Claim 1, the design of the installation for regulating the discharge from the reservoir is characterised by the temperature-controlled operation of the sluice-gates T by a floating body G.

4. In accordance with Claims 2 & 3, the design of the installation is characterised by the conduction of the heavy-water from sluice-gates T to the top of the barrage-wall K^1 by means of rising-pipes W.

5. In accordance with Claim 4, the installation is further characterised by the conduction of heavy-water to the valley-side of the barrage-wall K in horizontal troughs U^1, U^2 and U^3 at various heights above the base.

6. In accordance with Claim 3, the installation is characterised by the provision of a floating body G, constructed as a diving-bell with variable air-content, which can be raised or lowered.

7. In accordance with Claim 6, the installation is characterised by the incorporation of an open-ended, vertically adjustable pipe R in contact with the atmosphere.

8. In accordance with Claim 5 the installation is characterised by the connection of the individual diverter-pipes U^1, U^2 and U^3, for the conduction of heavy-water, to their common rising-pipe W via closable valves V^1 and V^2.

See Figure 23
Note: Figures referred to in patent text relate to those indicated in fig. 23 of this book.

Glossary

ABRASION: A process in which one material is caused to rub against another. Where one material is harder than the other, the softer will be reduced in size or smoothed by the removal of minute fragments. (See corrasion)

ANODE: An electrode carrying a positive charge, to which anions-, also electrons, are attracted.

BIOELECTRICISM: A higher, more ethereal form of electricity involved in electrical interactions in living systems and tissues. It is responsible for the healthy decomposition (not putrefaction) of formerly living matter and the subsequent transmutation of this into development-ripe raw material in consort with its counterpart - biomagnetism.

BIOMAGNETISM: A higher, more ethereal form of magnetism and the counterpart of bioelectricism. It is the form of magnetism responsible for uplift (both physical and spiritual), levitation and the generation of life-enhancing energies.

CAISSON: A floating metal canister, generally cylindrical in form. The one described in this book is closed at the top and open at the bottom, and is used to open and close the sluices of the reservoir. Open at both ends, it is more commonly used in bridge-building, to exclude water from the areas of the foundations, enabling their construction.

CARBONES: Principally those basic elements and raw materials of carbonous nature, although the term also includes all the elements of the chemist and physicist with the exclusion of oxygen and hydrogen. They are what Viktor Schauberger called "Mother-Substances", as they form the matrix from which all life is created. (See Footnote 6, p.20)

CAVITATING ACTION: Water has the capacity to dissolve matter and hold it in suspension. Used in this context - water making cavities in dam walls.

CENTRIFUGENCE: The function of so-called centrifugal force, which acts from the inside outwards. This is conventionally thought to eject any material exposed to it radially from the centre outwards, whereas in actual fact the material is expelled tangentially.

CENTRIPETENCE: The function of centripetal force. This is a force that acts from the outside inwards. Its most frequently observed manifestation takes the form of vortices.

CENTRIPULSER: A device having a number of whorl-pipes attached to a central hollow hub, whereby the medium (water or air) is moved in such a way that the forces of centrifugence and centripetence operate on a common axis. As the water is centrifuged from the centre of the hub outwards through the whorl-pipes, it is also caused to inwind centripetally due to the spiral configuration of the latter.

CORRASION: A process of mutual abrasion.

CYCLOID-SPIRAL-SPACE-CURVE MOTION: This can be a simple helical or spiral motion about the longitudinal axis, which on occasion pulsatingly expands from and contracts towards this axis. It can also embody a double spiral movement, in which the moving medium spirals about itself, while simultaneously following a spiral path. It is a form of motion analogous to the rotation of the Earth about the Sun, where the Earth gyrates about its own axis while moving along its orbital path. It is the form of motion Viktor Schauberger referred to as the "original" or "form-originating" motion responsible for the evolutionary dynamics of the Earth and Cosmos.

DENSATION: The process of becoming physically denser or more condensed.

DIELECTRIC VALUE: This refers to the capacity of a given substance to resist the transfer of an electric charge. The base value for a dielectric is that of a vacuum = 1. Water has one of the highest dielectric values, namely 81, which means that it is 81 times more resistant to the transfer of a charge than is a vacuum.

DYNAGENS: The entities or ethericities belonging to the 4th and 5th dimensions which enhance the creation of dynamic energy on lower planes of existence.

DYNAMIC ENERGY: This is energy that has more to do with the energising of all life-processes, subtle and otherwise, than purely physical phenomena for which the term kinetic energy, i.e. energy in motion, is normally used. (See potential energy)

DYNAMITIC SUBSTANCES: The violent, concentrated effect of oxygen in a spacially compressed, carbone-hungry form.

ELECTROZOIC ESSENCES: Also interpreted as animalistic or organismic essences (See p.43)

EMANATION: Any form of gaseous, vaporous, ethereal, spiritual, or electromagnetic emission of radiation, rays or energies.

ETHERIALISATION: The process of raising or exalting energies or matter to higher, more subtle states of being.

ETHERICITIES: This refers to those supra-normal, energetic, bio-electic, bio-magnetic, catalytic, high-frequency, vibratory, super-potent energies of quasi-material, quasi-etheric nature belonging to the 4th and 5th dimensions of being. (See Footnote 12, p.24)

FRUCTIGENS: The ethericities (subtle energies) responsible for increasing the fecundity or capacity for fructification and fertilisation of and by living things.

HALF-HYDROLOGICAL CYCLE: A truncated version of the full hydrological cycle in which no rainwater infiltrates the ground, but either drains away over the ground surface or re-evaporates into the atmosphere with unnatural rapidity, leading to excessive agglomerations and the uneven distribution of water vapour

HYDROLOGICAL CYCLE: The full, balanced and regulated natural cycle of water from deep within the Earth to the upper regions of the atmosphere and back, in which rainwater is able to percolate into the ground and the amount of atmospheric water is more evenly distributed and maintained at a more or less constant level. (See half-hydrological cycle)

IMMATURE WATER: Groundwater that has not yet accumulated and absorbed minerals, salts and trace-elements, which it requires in order to become mature.

IMPELLER: A mechanism for moving water or other liquid mechanically. Centrifugal impeller: the intake of water is along the axis of rotation in front of and perpendicular to the radially-ribbed impeller disc and is expelled tangentially under pressure at right-angles to the direction of inflow due to the action of centrifugal force. It has a disintegrative effect on water.

CENTRIPETAL IMPELLER: The water is introduced tangentially and exits axially in a longitudinal vortex down the central axis of rotation, which creates suction, cools and coheres the structure of the water.

INDIFFERENCE: Generally speaking, an unstable state of equilibrium where the organism or system in question is possessed of its highest potential, vitality, health and energy and is therefore able to operate at the optimal temperature and/or energy level appropriate to its proper function. Viktor Schauberger also defined this condition as "temperatureless". For human beings this state of indifference obtains at a temperature of +37° Celsius, and for water relates to its condition of least volume, highest density and energy content at a temperature of +4° Celsius, its so-called anomaly point.

INERTIA: The tendency or capacity of a given object or system to resist movement, acceleration or any change of status.

JUVENILE WATER: Akin to immature water, the term juvenile generally refers to rainwater, which lacks minerals, salts and trace-elements.

KINETIC ENERGY: Energy in motion or doing work. (See potential energy and dynamic energy)

LAMINAR FLOW: A condition in which the various strata of water within a given water-body flow without turbulence.

LAW OF ANTI-CONSERVATION OF ENERGY: The law postulated by Viktor Schauberger, where the amount of available energy, potential, dynamic or kinetic is not constant, which, by means of the appropriate device or dynamic process, can be increased at will to virtually any order of magnitude. It is the rational counterpart of the Law of Conservation of Energy.

LAW OF CEASELESS CYCLES: The primordial, immutable law of Nature that governs and is responsible for all cyclical phenomena such as the changing seasons, the alternation between night and day, the ebb and flood of tides, the diurnal fluctuations in the flow of sap in trees, the alternating pulsations between electric and magnetic fields, the movement of galaxies, and so on.

LAW OF COMMUNICATION: The law relating to liquids, which states that if any two or more bodies of a given liquid, water for instance, communicate directly with one another via some form of opening, then the surfaces of the respective liquids are brought to a common, uniform level, provided always that they have the same specific density or weight.

LAW OF CONSERVATION OF ENERGY: The law stating that the amount of energy throughout the Universe is finite; that there can neither be more nor less energy, which therefore always remains constant and thus can never be lost. Energy merely changes from one form to another, such as the transfer from a potential state to a kinetic state and vice versa.

LAW OF GRAVITY: The law governing the attraction of bodies towards the centre of a heavenly body or the mutual attraction between two or more such bodies. (See Law of Levity)

LAW OF LEVITY: The law postulated by Viktor Schauberger that governs and is responsible for all upward movement of energy, uplift, upward growth, the upright stature of human beings, animals and other organisms, and is the counterpart to the Law of Gravity. As the force of gravity decreases the force of levity increases.

LAW OF THERMODYNAMICS, SECOND: The law related to temperature derived from the Law of Conservation of Energy, stating inter alia that with no additional input of energy from some external source, the energy in all closed systems (the whole universe included) will eventually be transformed into heat and ultimately reduced to a condition of uniform temperature known as the 'Heat Death'.

LIGHT-INDUCED GROWTH: The rapid and unhealthy increase in the girth of shade-demanding species of timber when over-exposed to light, radiation and heart from the Sun.

NASCENT SPRING WATER: Immature water within the central stratum of the groundwater, having a temperature of about +4° Celsius.

NATURALESQUE: Refers to artificially contrived processes or mechanical devices that conform to or emulate Nature's laws, or operate in a naturally correct way.

POTENTIAL ENERGY: Stored energy or energy that as yet is unmanifested as dynamic or kinetic energy.

QUALIGENS: The ethericities responsible for the enhancement of quality and increase in quality matter.

SEEPAGE SPRING: A spring that is formed when percolating groundwater encounters an impervious stratum and drains away over the stratum surface under the influence of gravity towards the point of egress. The temperature of such springs generally conforms to the ambient ground temperature.

TEMPERAMENT: In Viktor Schauberger's terminology, this refers to the behaviour, character, gender and intrinsic properties, sometimes temperature-induced, of various immaterial and other energies, such as electricism, biomagnetism, gravity and levity as well as the media of earth, air and water.

TEMPERATURE GRADIENTS: In terms of Viktor Schauberger's concepts, temperature gradients are principally related to the direction of movement of temperature within and between the respective temperatures of the ground, water and atmosphere, which can either take a positive or negative form. A positive temperature gradient occurs when the direction of temperature movement is towards the anomaly point of water, i.e. towards +4∫ Celsius. A negative temperature gradient occurs when the direction of temperature movement is either upwards or downwards from +4°Celsius.

TRACTIVE FORCE: The force that acts to 'shear off' or to dredge and dislodge sediment. (See Footnote 3, p.16)

TRIBO-LUMINESCENCE: An internal glow or luminescence produced when two or more crystalline rocks of similar composition are rubbed hard together or struck against one another and is attributed to the energy given off by the electrons contained the rocks as they return from a pressure-induced, excited state to their rest orbits. As a phenomenon it can occur both in air and under water. (See Footnote 18, p.39)

TURBIDITY: A measure of the opaqueness, cloudiness or muddiness of water due its content of suspended matter.

VALENCY: Highly polarised, pregnant state.

WATER MASSES: Both the body of water generally, but also the various swirling volumes and filaments of water of different temperatures, densities and energetic content whose values are prescribed by the inner densities of the water.

WHORL-PIPES: Pipes, principally made of copper or its alloys, having a spiral configuration akin to that of a Kudu antelope, through which the transported medium is caused to move centripetally and vortically in a double spiral motion.

Index

agriculture & groundwater, 104
agriculture, declining, 140
AIDS, 5
aircraft, powering, 156
air eggs, 24
Amazon, 145
annual rings, 36
anode functions, 42
anode-water+, 72
anomaly point, 28
aquatic plants, behaviour of, 143, 148
aquifers, 50
Arctic Ocean, 114, 175
aridity, increasing, 94
Atlantis, 136
atmosphere, 66
atmospheric cycle, 135, 145
atmospheric discharge, 161
atmospheric pressure, drop in, 109

bacteria, 161
bacteria, high-quality, 49
bacteria, water-polluting, 63
Bergel, Kurt, 62
bipolar energies, 161
blood capillaries, 56
blood decomposition, 28
blood transfusions, danger of, 54
blood, composition of, 49
blood, de-energised, 26
boar, pissing, 83
Bodensee, 180
bore-hole, 58
boundary layer, 74
brake-curves, 84
brake-groins, 210
brakes, hydraulic, 163
Buchinger, Rudolf, 190

calcium-bicarbonate, 207
cancer, 8, 19, 47, 53

capillaries, 17
carbon-dioxide, 53, 59, 66, 72, 137, 205
carbon-dioxide, colder, 25
carbones, 19, 40, 49, 72,75, 172, 196
carbones, deep sea, 58
cabonesphere, 66
carbone-spherics, 73
carbonic acid, 54, 65, 72, 141, 173, 182
Carthage, 140
cathode functions, 42
cathode-water, 72
cavitating action, 123
cellulose, 198
central axis of flow, 109, 111
central axis oscillation, 116
channel profile, 110, 116
Chinese religion, 68
coal, 74, 75, 77
cold water core, 117
combustion phenomena, 49
conduits, natural stone, 53
core-water body, 109, 168
crusher dust, 12
cycloid motion, 27
cycloid space-curve, 84
cycloid space-curve motion, 89

dam building, 101
dam design, 121
dam discharge device, 212
dam walls, mass-concrete, 122
dam walls, sealing, 213
dams, 177
Danube, 16, 38, 135, 144, 170
Danube/Inn confluence, 86
delta formation, 175
deserts, 147
Dog Spring, 20
dredging, 179
drinking water, production, 207

drinking water, sterilisation, 45
drought, 139
dynagens, 24, 43, 163
dynamitic substances, 86

Earth's planetary motion, 35
Earth breathing, 196
earth eggs, 10, 24
economic collapse, 159
economic decline, 147
egg shapes, 182
Egyptians, ancient, 127
Ehrenberger, Dr, 170, 176, 183, 193
electricity generation, 38. 208
electro-chemistry, 78
energies, radiant, 75
energies, recycling, 161
energiewasser, 118
energy bodies, 179, 180
energy cycle, 66
energy-form changes, 147
energy-form, electromagnetic, 138
energy-form, electrostatic, 138
energy-form, kinetic/potential, 138
energy-form, turbulent, 146
ethericities, 24, 160
Etsch, 16, 145, 171
evolution, forms of, 67
Exner, Prof Wilhelm, 81, 84. 176

fertilisation processes, 161
fertilisation, artificial, 105
field, wrongly-ploughed, 166
fish biting, 153
fish spawning, 152
floods, 94, 115, 151, 177
flow-axis, cooling, 109, 168
fluctuations, diurnal, 147
fluting, rifled, 202
flying machines, 26

Forchheimer, Prof Philipp, 15, 81, 94, 106, 117, 131, 134, 189, 193
forest, clear-felling, 140, 157
+4°C centre stratum, 5, 96, 197
fructigenic matter, 89

Ganges flood-plain, 157
Garonne, 16
grazing, decline in, 104
geostrophic effect, 112, 115
germination zone, 28
Ghijbens, Badon, 57
Gobi desert, 147
Goethe, 89
Gold of the Nibelungs, 16
gravitation, 26
ground drainage, 101
groundwater, 67
groundwater and agriculture, 104
groundwater drainage, 99
groundwater reservoirs, 102
groundwater retreat of, 104
groundwater table, 49, 74, 96
growth, light-induced, 36, 140
Gruppe der Neuen, 22
guide-vanes, 202, 204
Gulf Stream, 164

H_2O, 195
haff formation, 175
Hahn, Otto, 36
Hasenöhrl-Einstein equation, 3
Hauska, Prof, 189
heart, as pump, 62
heat *vs* motion, 181
helical motion, 61, 63
hetero-polar, 178
Hitler, Adolph, 87, 191
Hohl, Arnold, 22
Hörbiger, 198
humanity, decadence of, 68
hunger, world, 26
hydraulics, 135
hydro-electric device, 208
hydro-electric power, 141, 157
hydro-electric turbines, 155
hydrogen, 40, 163, 198
hydrogen, organic, 182
hydrological cycle, full, 95. 101
hydrological cycle, half, 96, 101
hydrolytic transformation, 181

ice crystals, 198
immune system, 4
Implosion, 13, 26, 62, 81
ionisation, 178

iron ochre, ore, 54
iron oxide, 66

Karlsbad hot spring, 17, 107
karst development, 158
kidney disease, 26
Kienböck, 190
Köber, 193
Kokaly, Aloys, 26

lakes, impounded, 102, 119
levitation, 26
living standards, 27
log-flumes, 81

machines, elemental, 164
marinectic lakes, 58
Mark, Prof Dr, 184
matter, deposition of, 147
Mensch und Technik, 22, 42, 44, 159
Mississippi, 16
monoculture, 36
moss, 112
motion, centripetal, 152
motion, laminar, 106, 138
motion, natural, 34
motion, perpetual, 43
motion, torsional, 203
motion, turbulent, 106, 138
mountain spring principle, 75

Neuberg log flume, 82, 107
Nile, 175
nitrogen, 198
nutrients, decrease of, 143
nutrients, distribution of, 141

oil, 20
organic loops, 23
organic machines, 163
Our Senseless Toil, 2, 15, 45
over-trickling, 126, 212
oxygen, 27, 28, 40, 72
oxygen, aggressive, 7, 36, 48, 57, 65, 73, 89, 173, 197
oxygen, colder, 25
oxygen, condensed/crystallised, 198
oxygen, concentration of, 21
oxygenes, 49
oxygen-spherics, 73

paths, multiple helical, 204
phos-substances, 162
pipe capillaries, 65
pipe walls, 201
pipe, double-spiral flow, 63

pipe, double-torsional flow, 185
pipelines, transport by, 48
Planck, Max, 87
Plato, 136
Po, 114, 145, 171
polar forces, 40
Pöpel, Franz, 13, 38
pot-holes, 149
protein, 23
Pröckner, Richard, 187
pumps, water, 70
pyrite, 76

qualigen, 27

rainwater temperature, 94
Rappolt, Otto, 144, 171
Renault, Gerard, 184
reservoir, properly-constructed, 103
reservoir, subterranean, 96
Rhine, 38, 136, 144, 171, 176, 179
Rhinegold, 16
river bank rectification, 102
river bend formation, 107, 111
river channel, 44
river heat creation, 27
river regulation, 115
river regulation device, 209
riverbed formation, 111
riverbed gradient, 92, 148
Romans, 53, 71
rust, 53

salts, bicarbonates of, 76
sap in trees, 62
Sarkar, Dagmar, Dr, 165
Schaffernak, Prof, 86, 175, 189
Schauberger, Walter, 13
Schocklitz, Prof, 86, 175
science, tenets of, 194
sediment deposition of,112
sediment, transport of, 167
sedimentating action, 123
Serbia, 171
silver, 182
Smörcek, Prof, 86, 189
snail, blood-vessels of, 61
snow melt-water, 78
soil fertility, decline in, 104
solar irradiation, excessive, 35
springs, 38, 141, 149
springs, high altitude, 75, 90, 97
springs, hot, 98
springs, in summer, 100

springwater, 25, 49, 72
springwater production, 205
starches, 198
Steyrling, 180
stones, metalliferous, 178
storms, cyclonic, 139
stratosphere, 66
streams, cold, affluent, 145
suction forces, 168
sugars, 198
Sun, 193, 195
Sun's light, shielded, 145
Sun, cold, 102
swamp development, 104
Sweden, 112

Tagliamento, 16, 145, 171
tar, pipes, application in, 54
Tau, 38, 39, 167, 176, 179
temperature change, 76
temperature gradient, negative, 54, 95, 108
temperature gradient, positive, 95, 98, 108
temperature gradient, re-establishment of, 105
temperature gradient, regulation, 94, 115, 119
temperature gradient, reversal, 120, 155, 174
temperature gradient, reverse, 155
temperature gradient, summer/winter, 78
temperature reversal, 174
Tepl, 17
Tepl dam, 131
Tepl /Karlsbad confluence, 107
Tepl/Eger confluence, 18, 111, 133
Tepl/hot spring confluence, 111
Thales of Miletus, 15, 136
Thermodynamics, Second Law of, 13
timber floatation, 109
timber, rafting of, 156
torrent confinement device, 209
torrente, 113
tractive force,16, 108, 161
trees & water, 25
trees, cancerous diseases, 200
trees, shade-demanding, 36
triboluminescence, 17
trout 89ff, 149, 154

turbine blades, corrosion, 57
turbines, 54
turbulence, 81, 89, 154
turbulence phenomena, 106

unemployed, 159
ur-feminine/masculine, 27
ur-forces, 165
U-tube experiment, 50

vacuum, 24
vegetation, build-up, 198
vegetation, change of, 76
vegetation, degeneration of, 19
vegetation, evolution of, 137
venereal diseases, 55
Venetian rivers, 113
Vidrasku, Prof, 144, 171
Vienna, 53
vortex-creating elements, 203
vortices, counter-, 155
vortices, formation of, 149
vortices, peripheral, 151
vortices, train of, 116

Waldecker Dam, 123
Wasserwirtschaft Die, 15, 81, 94, 106, 135, 145, 187, 193
water balance, disruption of, 68
water brake, 209
water cycle, cause of, 85
water eggs, 22
water masses, 16
water masses decelerate, 112
water masses, distribution, 146
water masses, heavy, 112
water movement, centripetal, 152
water movement, positive/negative, 171
water movement, stratified, 106
water pipes. iron, 53
water pipes, wooden, 53
water quality improvement, 12
water storage, 9
water strata, core, 150
water stratification, 89, 106
water supply system, 53
water temperature, influence of, 89
water vapour, 142
water velocity, 92
water, and Sun, 34, 112
water, anomaly point, 5, 23

water, as a carrier, 172
water, as accumulator, 42
water, as Nature's housekeeper, 173
water, as radiant emission, 26
water, atmospheric, 25, 72
water, boiling point of, 25
water, carbone-deficient, 49
water, chlorination of, 4, 47
water, conduction, 201, 203
water, decomposition of, 69
water, de-energised, 26
water, deep-sea, 59
water, deterioration of, 45
water, disappearance of, 68
water, distilled, 49
water, energy in, 22
water, energy-form, 137
water, evaporation of, 91
water, explosions in, 163
water, holes, 11
water, formula for, 193
water, freezing point of, 25
water, growth of, 181, 185
water, healthy drinking, 37
water, heavy, 41, 212
water, heavy/light, 214
water, high-frequency, 26, 29
water, high-grade apparatus, 79
water, irradiation, 47
water, mature, 71
water, medicinal, 76
water, new, formation of, 85, 163, 186
water, oxygen-rich, 78
water, pulsation of, 66
water, radioactive, 77
water, ripe, 72
water, specific heat, 142
water, subterranean, 91
water, varieties of, 3
water, winter, 25
watercourse in sunlight, 112
waterworm, 20
wells, 11, 55
Weyrauch, Robert, 108, 122, 176
Wiener Tag, Der, 193
Wiluhn, Minister, 191
Woltmann-vane, 173

X-ray radiation, 41

Zimmermann, Prof Werner, 38, 176, 179, 186